AF702565

Breaking Feminist Waves

Series Editors:

LINDA MARTÍN ALCOFF, Hunter College and CUNY Graduate Center
ALISON STONE, Lancaster University
GILLIAN HOWIE†, University of Liverpool

For the last twenty years, feminist theory has been presented as a series of ascending waves. This picture has had the effect of constraining the way we understand and frame new work as well as deemphasizing the diversity of past scholarship. The aim of this series is to attract original scholars who will unearth neglected contributions to feminist theory and offer unique interpretations of past scholarship. By breaking free from the constraints of the image of waves, this series will be able to provide a wider forum for dialogue and engage historical and interdisciplinary work to open up feminist theory to new audiences and markets.

LINDA MARTÍN ALCOFF is Professor of Philosophy at Hunter College and the City University of New York Graduate Center. Her books include *Visible Identities: Race, Gender and the Self*; *The Blackwell Guide to Feminist Philosophy* (coedited with Eva Kittay); *Identity Politics Reconsidered* (coedited with Moya, Mohanty, and Hames-Garcia); and *Singing in the Fire: Tales of Women in Philosophy.*

ALISON STONE is Professor of European Philosophy at Lancaster University, UK. She is the author of *Petrified Intelligence: Nature in Hegel's Philosophy*; *Luce Irigaray and the Philosophy of Sexual Difference*; *An Introduction to Feminist Philosophy*; and *Feminism, Psychoanalysis, and Maternal Subjectivity*; and the editor of *The Edinburgh Critical History of Nineteenth-Century Philosophy.*

Titles to date:

Unassimilable Feminisms: Reappraising Feminist, Womanist, and Mestiza *Identity Politics*
by Laura Gillman

Further Adventures of The Dialectic of Sex: *Critical Essays on Shulamith Firestone*
edited by Mandy Merck and Stella Sandford

Hegel's Philosophy and Feminist Thought: Beyond Antigone?
edited by Kimberly Hutchings and Tuija Pulkkinen

Femme*nism and the Mexican Woman Intellectual from Sor Juana to Poniatowska: Boob Lit*
by Emily Hind

Between Feminism and Materialism: A Question of Method
by Gillian Howie

Resonances of Slavery in Race/Gender Relations: Shadow at the Heart of American Politics
by Jane Flax

The Many Dimensions of Chinese Feminism
by Ya-chen Chen

Rousseau in Drag: Deconstructing Gender
by Rosanne Terese Kennedy

Undutiful Daughters: New Directions in Feminist Thought and Practice
edited by Henriette Gunkel, Chrysanthi Nigianni, and Fanny Söderbäck

A Theory of Freedom: Feminism and the Social Contract
by Shay Welch

Theory on the Edge: Irish Studies and the Politics of Sexual Difference
edited by Noreen Giffney and Margrit Shildrick

Gendered Readings of Change: A Feminist-Pragmatist Approach
by Clara Fischer

Feminism, Time, and Nonlinear History: A Polytemporal Approach
by Victoria Browne

Socrates and Diotima: Sexuality, Religion, and the Nature of Divinity
by Andrea Nye

Socrates and Diotima

Sexuality, Religion, and the Nature of Divinity

Andrea Nye

palgrave
macmillan

SOCRATES AND DIOTIMA

Copyright © Andrea Nye, 2015.

All rights reserved.

First published in 2015 by
PALGRAVE MACMILLAN®
in the United States—a division of St. Martin's Press LLC,
175 Fifth Avenue, New York, NY 10010.

Where this book is distributed in the UK, Europe and the rest of the world, this is by Palgrave Macmillan, a division of Macmillan Publishers Limited, registered in England, company number 785998, of Houndmills, Basingstoke, Hampshire RG21 6XS.

Palgrave Macmillan is the global academic imprint of the above companies and has companies and representatives throughout the world.

Palgrave® and Macmillan® are registered trademarks in the United States, the United Kingdom, Europe and other countries.

ISBN: 978–1–137–51601–5

Library of Congress Cataloging-in-Publication Data

Nye, Andrea, 1939–
Socrates and Diotima : sexuality, religion, and the nature of divinity / Andrea Nye.
pages cm.—(Breaking feminist waves)
Includes bibliographical references and index.
ISBN 978–1–137–51601–5 (hardback : alk. paper)
1. Love. 2. Socrates. 3. Symposium (Classical Greek drinking party) 4. Plato. Symposium. I. Title.

BD436.N94 2015
184—dc23 2015003951

A catalogue record of the book is available from the British Library.

Design by Newgen Knowledge Works (P) Ltd., Chennai, India.

First edition: August 2015

10 9 8 7 6 5 4 3 2 1

Contents

Series Foreword

Breaking Feminist Waves is a series designed to rethink the conventional models of what feminism is today, its past and future trajectories, moving away from the metaphor of waves. For more than a quarter of a century, feminist theory has been presented as a series of ascending waves, imagery that constrains the way we understand what feminism has been and where feminist thought has appeared. This imagery simplifies the rich and nuanced political and philosophical diversity that has been characteristic of feminism throughout, and, most disturbingly, it restricts the way we understand and frame new work. The aim of this series is to rethink the history and actuality of feminist theory outside of these restricting metaphors.

This series provides a forum to reassess established constructions of feminism and of feminist theory. It provides a starting point to redefine feminism as a configuration of intersecting movements and concerns, with political commitment but, perhaps, without a singular center or primary track. The generational divisions among women do not actually correlate to common interpretive frameworks shaped by shared historical circumstances, but rather to a diverse set of arguments, problems, and interests affected by differing historical contexts and locations. Often excluded from cultural access to dominant modes of communication and dissemination, feminisms have never been uniform nor yet in a comprehensive conversation. The generational division, then, cannot represent the dominant divide within feminism, nor a division between essentially coherent moments; there are always multiple conflicts and contradictions, as well as differences about the goals, strategies, founding concepts, and starting premises. In particular, this series provides a space for exploring the sometimes surprising philosophical and theoretical resources that feminists have taken as their starting premises at different times and in varied cultural contexts.

In the contemporary world, the problems facing women, feminists, and feminisms are as acute and pressing today as ever. Featuring a variety of disciplinary and theoretical perspectives, *Breaking Feminist*

Waves provides a forum for comparative, historical, and interdisciplinary work, with special attention to the problems of cultural differences, language and representation, embodiment, rights, violence, sexual economies, and political action. By rethinking feminisms' history as well as their present, and by unearthing neglected contributions to feminist theory, this series intends to unlock conversations between feminists and feminisms and to open up feminist theory and practice to new audiences.

LINDA MARTÍN ALCOFF
and
ALISON STONE

Prologue

> *A human being has roots by virtue of his real, active, and natural participation in the life of a community, which preserves in living shape certain particular treasures of the past and certain particular expectations for the future.*
>
> —*Simone Weil,* The Need for Roots[1]

Few women's voices survive from classical antiquity. What we know of the nine-volume oeuvre of the Aeolian poet Sappho, universally praised as one of the greatest of Greek poets, is only a few passages quoted by classical writers and some fragments unearthed from archeological excavations. All that remains of the thought and teaching of Aspasia, companion and adviser to Pericles, are bits and pieces of wisdom reported by contemporaries, denunciations from conservative critics of democracy, and a model speech repeated by Socrates in Plato's dialogue the *Menexenus.* One or two letters purporting to be by women followers of Pythagoras and some collaborative commentaries attributed to the murdered Neoplatonist philosopher Hypatia make their way into anthologies but little else.

There is no doubt, especially in the area of religion, that women were active and influential in Greek culture. Inscriptions and records document the many spiritual and civic functions of priestesses who officiated in state and local cults with all the authority and respect given to priests. At the same time, no sustained account is available from female religious officials themselves of the meaning of these rites and rituals. Nor do we have participant views of religious festivals and celebrations attended exclusively by women such as the Thesmophoria in honor of Demeter and maenadic rites related to Dionysus, but must rely on male imaginings as to what might have gone on at these all-female gatherings. The result for women scholars looking for "treasures of the past" and "expectations for the future" can be a permanent state of opposition, a sense of being "without roots" in the life of an intellectual community.

One day, reading yet again one of the seminal texts of the Western tradition, I found myself paying closer attention. In his dialogue the *Symposium*, Plato tells the story of a drinking party at which Socrates and some of his friends entertain each other with speeches in praise of love. Various speakers stand up and take the floor and speak eloquently. Love makes a man courageous in battle, said one, as he tries to impress his lover. Yes, said another, but there must be advantages for both in the relationship. Another guest, a medical doctor, reminded the company of considerations of hygiene. The comic playwright Aristophanes told an amusing folktale. But with Socrates comes a change in tone. When it is his turn, he does not speak in his own voice. Instead he recounts word for word lessons in love he once had from a woman, as he puts it, "wise in matters of love and in much else." It was she, he said, who taught him everything he knows and still believes about love and divinity.

As I read the famous passages that follow, disputes about sexuality, love, and religion were everywhere. Evangelical Christians tried to block sex education in the schools and stop the distribution of condoms. Gay rights advocates proclaimed a right to marry. Feminists insisted that the double standard in sexual morality be abandoned and defended the right to abortion. Conservatives deplored a breakdown in traditional family values. Sexuality was a battlefield with casualities on both sides, as women lined up against men, atheists against Christians, liberals against conservatives. All were armed with voices from the past proclaiming, on the one hand, the weakness of the flesh and the evils of fornication and, on the other, the truth of science and the sanctity of individual rights. This time, however, as I followed Socrates's recreation of his lessons in love from a woman "honored by the gods," checking the Greek text along with several of the many English translations, I was hearing another voice, not the provocatively ironical Socrates or the visionary Plato, not a pretentious priest like Euthyphro or a self-important Sophist like Hippias, but a woman's voice speaking with some authority.

I was on precarious ground. The standard view is that Socrates's mentor in love is a fiction created by either Socrates or Plato for reasons no one is able to explain with much certainty. But I was hearing ideas expressed that were neither Socratic nor Platonic, and I was remembering related ideas from other texts, from philosophical poems by Parmenides and Empedocles, from remnants of Sappho's lyrics and choral odes from tragic dramas, ideas not highlighted or noted in standard commentaries. The notice Diotima had received from feminists was mostly negative. Liberal philosophers like Martha Nussbaum dismissed her as no more than a stand-in for Plato. Postmodern theorist Luce Irigaray approved her positive account of eros, but immediately chastised her for failing "in method" the minute she began talking about immortality.

But what if, I asked myself, Socrates's lessons in love were not pure invention and not poorly articulated Platonism; what if, like other points of view expressed in the dialogues, they represent an important strand or undercurrent in the rich mix of ideas and theories debated and discussed in fifth-century Athens? What if Diotima's references to Aphrodite, Moira, and Eileithyia were not rhetorical flourish but references to sacred teaching associated with those divinities? In short, what if Diotima's lessons in love were exactly what they purport to be: religious teaching that places eros at the very heart of spiritual life as a privileged access to virtue and immortality?

What follows is an attempt to answer these questions. What I explore is not only what might have been the original religious meaning of Diotima's teaching, but also the story of how that meaning was lost. In Part I, I return to the source and read Socrates's account of his sessions with Diotima in the context of what we know of the historical Socrates and of Greek religious practice. In Part II, I show how the message of Diotima's lessons in love was progressively obscured and misappropriated in Plato's later writings, at the Academy he founded in Athens and in the writing of third-century "neoplatonists" like Plotinus. I go on to trace the influence of Platonist doctrine in Christian thought and the demonization of the daemonic eros that in Diotima's teaching is a pathway to divinity. Part III returns to find undercurrents of an older suppressed tradition of thought in Platonic and Christian texts as well as in contemporary liberation and feminist theologies. What results, I hope, is a view of divinity and spirituality consistent with the findings of modern science, effective in inspiring virtuous action, welcoming to both women and men, and a surer, more realistic source of consolation in the face of mortality and death than any Platonic heaven of winged souls.

If there is any truth in my interpretation of Diotima's lessons in love, it must be taken for what it is, a fresh reading of an old text by a feminist philosopher looking for answers to some pressing contemporary questions. Have there been, are there still, forms of religious devotion that do not undermine women's power and spiritual authority? Is there a noncredulous form of religious belief consistent with the aims and values of empirical science? Most important, is it possible for women and men to find common ground on the contested battlefields of sexuality and religion? Although there can be no final proof that a Mantinean priestess named Diotima visited Athens, was sought out by Socrates, met with him on several sessions, and mentored him in "matters of love," nor is there any certain proof that she or the views she expresses are fictions. What is more important than the existence of any particular named individual is an understanding of eros and spirituality distinctively different both from what later emerges as Platonism and from the biblical traditions that along with Platonism are taken as the bedrock of Western thought. Writing in the aftermath of World War II, Simone Weil wrote of the "uprootedness" that results from violent conflict. Perhaps an analogy might be made with our postmodern, poststructuralist age of disbelief. If culture wars do not tear up the physical earth, they tear away common threads of memory and heritage that create a human community. But Weil also wrote of the process of recovering and regrowing intellectual roots that can support a place to stand and a vision for the future.

Introduction

There has been much debate about how to read and interpret the many ideas and opinions expressed in Plato's dialogues. One approach, currently popular, is to treat the conversations recreated in the dialogues as literary fiction. On this view, characters and settings in the dialogues were invented or contrived by Plato to showcase or try out positions of his own, either programmatically or at different stages in his thinking.[1] Another approach is to categorize "early dialogues" as "Socratic"—more or less accurately depicting the historical Socrates in conversation with his contemporaries—but to read "middle" and "late" dialogues as "Platonic," with Socrates increasingly a mouthpiece for Plato's own evolving theory of Forms. Neither approach is without difficulty. Given the diversity of views expressed in the dialogues and the many accurate and specific references to historical persons and historical events, the fictional thesis is hard to sustain. Not only are many of the characters public figures, but some of them were Plato's own relatives, including his mother's uncle Critias, his mother's brother Charmides, and his brothers Adeimantus and Glaucon who figure prominently in the *Republic.* Nor has it been easy to find any clear dividing line between so-called early Socratic dialogues and dialogues in which Plato supposedly expresses his own ideas. Adding to the difficulty is the lack of reliable evidence as to when each dialogue was written, the evolution of Plato's thinking, or events in his life.[2]

In making a division between "Socratic" and "Platonic" dialogues, the *Symposium* presents particular difficulties. In that dialogue, Socrates proclaims that he learned everything he knows about love from Diotima and that he remains convinced by her teaching. At the same time, Diotima's teaching is not consistent with approaches traditionally classed as "Socratic." Instead of engaging in Socratic dialectic, submitting all claims to knowledge to critique, Diotima propounds a doctrine. In place of the Socratic "know yourself," she urges Socrates to look around and observe the facts of natural life. But nor is Diotima's teaching consistent with what is traditionally classed as "Platonic." There is no escape from bodily existence, no rejection of animal appetite, no immortal souls winging their way to heaven to view eternal Forms of Goodness and Justice.

Nevertheless, although there has been willingness to grant the historicity of many if not most of the persons and ideas in the dialogues, in the case of Diotima the supposition of authenticity has been particularly hard to accept. Yes, one might imagine Socrates taking on the venerable and celebrated philosopher Parmenides, or exposing the shallowness of a Sophist like Hippias, or ridiculing the self-righteousness of do-gooders like Euthyphro, but taking lessons from a woman and insisting that he learned everything he knows and still believes about love from her has strained scholarly credulity. What is the great Socrates doing repeating the teaching of a woman? And why would Plato make a point of recounting such a teaching? Is this only a playful move in a game of seduction with Socrates cleverly hoping to soften his "elenchus" of the coveted Agathon with a confession that he too was once in need of instruction? Is Plato artfully clothing a new vision of masculine virtue in feminine dress? Is Plato on the way to his theory of Forms, but not yet there? Or is this no more than a clever way for Plato to talk about Socrates's initiation into mysteries of eroticism without the implication that there had been pedophilic sexual contact between Socrates and a male mentor? Many and endlessly creative have been efforts to explain away Diotima's teaching as an invention on the part of either Socrates or Plato. Scholars struggle, identifying Diotima's teaching as Platonism not yet fully developed" or "tailored for a particular audience" or, in the end, as simply bizarre, absurd, an unfortunate lapse on Plato's part.[3]

Much of the debate passes over what Plato's dialogues purport to recreate: conversations between Socrates and his contemporaries taken from oral memories and written memoranda that were passed around among Socrates's followers in the difficult years before and after his trial and execution. The cast of characters is small. Almost all are historical figures, identifiable as from a few Athenian hereditary demes and related by blood or marriage. All were people that Plato knew or would have known about. With a few minor exceptions, historical references to events and dates are accurate and consistent. As Debra Nails puts it in her detailed "prosography" or collective biography of the "people of Plato":

> Plato did not invent Athenians with names, demes, and kin; he wrote about real people—some of them still alive and active and living in

> Athens—people with reputations, families, neighbors, and political affiliations, people who show up elsewhere. (*People of Plato*, xxxvii)[4]

Not only are the people in the dialogues known personages, but the religious, ethical, metaphysical views they express are regularly taken as indicative of existing schools of thought such as Pythagoreanism, Sophism, Ionian natural philosophy, and Orphism.

The telling and retelling of stories about the elusive and charismatic Socrates conversing with figures of his time provides the explicit narrative frame for most of the dialogues. Plato himself was too young to have been present at the conversations he recreates, but he was close enough to Socrates in the philosopher's last years for his absence at Socrates's deathbed to be noted and attributed to illness.[5] At Socrates's trial, he was present and could have taken notes from which he reconstructed Socrates's defense speech in the *Apology*. For dialogues in which Socrates is narrator, Plato could have heard an account of the conversation from Socrates himself. In other cases, he might have appealed directly or indirectly to participants to confirm what was said. The *Symposium* is a typical example.

As the dialogue opens, an unnamed person, perhaps Plato himself, is asking Apollodorus about a drinking party or "symposium" at which Socrates made a memorable speech in praise of love. Apollodorus responds that he has had many such requests: "Oh, if that's what you want to know about, it hasn't been very long since I had occasion to refresh my memory." Only a few days before, he goes on to explain, he had been asked to tell the story to Glaucon, who heard an incomplete version and wanted clarification. In fact, says Apollodorus, by now he has had to retell the story so many times that he has it down "pretty much by heart." Furthermore, he took time to seek out Socrates and confirm the details. After further urging on the part of his companion and some talk about the importance or lack of importance of philosophy, Apollodorus settles down to relate what was said at a drinking party at which Socrates recalled his lessons in love from a woman "honored by the gods."

In a hyper-literate age, descriptions of hearing, remembering, making notes of, and retelling stories might seem no more than literary conceit, but fifth-century Greece was still a largely oral culture. Skill in memorizing and recalling myths, poetry, speeches, and histories, sometimes with the help of written notes, is noted many times in the dialogues. In the *Theaetetus*, Euclides recalls a story that he remembered just after parting from the wounded warrior Theaetetus, a story told to him by Socrates about a conversation Socrates had with Theaetetus when Theaetetus was only a youth, Terpsion, eager to hear such a story, asks whether Euclides might be able to repeat it. Euclides responds:

> Of course not just from memory, but I made some notes at the time as soon as I got home. And later on I wrote out what I could recall at my leisure. Then whenever I was in Athens, I would question Socrates upon any detail where my memory failed, and I made corrections when I returned. In this way I have pretty much the whole conversation down in writing. (143a)

Terpsion says he has heard about such notes, and Euclides calls for a servant to bring out the manuscript. Giving it to Terpsion to read, he calls the young man's attention to the style of the writing.

> You see how I wrote the conversation—not in narrative form, as I heard it from Socrates, but as a dialogue between him and the other persons he told me took part…I wanted to avoid in the written account the tiresome effect of bits of narrative interrupting the dialogue, such as "and I said" or "and I remarked" whenever Socrates was speaking of himself, so I left out everything of that sort and wrote it as a direct conversation between the actual speakers. (143b–c)

In fifth-century Athens, there was nothing unusual about such feats of memory. Storytellers recited long sections of the *Iliad* from memory. Noteworthy orations were memorized and repeated verbatim.[6] At the heart of a young man's education was rhetoric, a prominent part of which was the art of remembering long passages from speeches and poems.

Another example occurs in the *Timaeus* with Critias's story of ancient Athenians. The day before Socrates had entertained the group with his elaborate description of a fictional ideal "republic." Now, he and his friends gather for a second day of "philosophical" amusement, and it is someone else's turn to "perform." Socrates calls for a different kind of story, not of a fictional republic, but an actual state in action, a state that once existed. Critias takes the floor prepared, he says, to tell a true story, a story that originated with Solon, a story he first heard from his grandfather and that he spent the night before recalling. He explains the process. He first heard the story when he was ten years old at his grandfather's knee, and even then he took pains to remember it exactly. "I listened with childlike interest to the old man's narrative; he was very ready to teach me, and I asked him again and again to repeat his words, so that, like an indelible picture they were branded into my mind" (26b–c). The day after he rehearsed the story several times with young friends to be sure he remembered all the details. In the same way, as represented by Plato in the dialogues, stories about Socrates, repeated and passed around in the anxious years before and after Socrates's death, were "branded into the memories" of the young men who gathered around him.

There is no doubt that Plato had his own interests in working memories of these conversations into a unique literary form. According to Aristotle, he took an early interest in philosophy.[7] He was familiar with Heraclitus's vision of a natural world in constant change. He was personally acquainted in Athens with Heraclitus's disciple Cratylus. He wondered how knowledge could be possible of a natural world that is constantly in motion and took note of the uncertain results of Socrates's search for universal definitions of moral qualities like justice or goodness. At some point in his thinking, said Aristotle, Plato gave up the possibility of any true knowledge of goodness in the natural world and turned to contemplation of a supersensual world of ideas, thus distinguishing himself from Socrates. That most of the conversations Plato chose to depict in the dialogues involve debates surrounding these issues is not surprising, but it does not show that persons involved in these debates or the points of view they express were Plato's inventions.

If there can be no hard truth in memoranda of long past conversations—no certainty that the young Theaetetus was really as precocious in his answers to Socrates's questions as Socrates remembered, or that Critias repeated accurately his grandfather's tale of ancient Athenians, or that a visiting priestess named Diotima instructed the young Socrates in exactly the terms recalled by Socrates and passed on by Apollodorus—there is something better. What Plato's dialogues preserve is not doctrine or theory, dead words that cannot answer back, but ideas sown in fertile minds where thoughts, as Socrates puts it in the *Phaedrus*, "contain seeds from which other ideas can grow up" (277a1–2).

* * *

Along with Aspasia, the advisor and lover of Pericles, and Sappho whose poetry was praised by Socrates, Diotima is one of the few women whose ideas are represented in the dialogues. Aspasia and Sappho are historical figures, well documented from other sources; the lack of such sources in the case of Diotima makes her existence less certain. According to Socrates, she was from Mantinea, an

Arcadian city known for its many temples, oracular sites, and religious traditions. Remote and rugged, the region resisted the Dorian invasions and kept many of its ancient traditions. Its population spoke an Arcado-Cypriot dialect closer to Mycenaean Linear B than to classical Greek, and its windswept shrines were popular with those in spiritual or medical trouble. At the time of his lessons, according to Socrates, Diotima was an official guest in Athens, invited by the democratic leader Pericles to give spiritual guidance in the face of a deadly plague that was spreading north toward the city.[8]

There was nothing unusual about such a summons. Relying on religious authority in times of trouble had many precedents. In Homer's *Iliad*, with Troy under siege by the Achaeans, the Trojan queen Hecuba calls on the priestess Theano to make offerings and prayers to save the city (*Iliad V* 297–311). Herodotus relates how in the face of Persian attack a priestess was the only one able to persuade the frightened citizens of Athens to evacuate the city and survive to defeat the invaders. In the *Phaedrus*, Socrates praised the role of Greek priestesses in conciliation and peace making: "When they were divinely inspired the Prophetess at Delphi and the priestesses at Dodona achieved much for which both states and individuals in Greece are grateful" (244b). In the *Meno*, he acknowledged "wisdom" he had acquired from "priests and priestesses of the sort who are able to give an account of the functions they perform" (81a). Women kept the keys to many temples and sanctuaries. They officiated at festivals. They supervised young girls who served for periods of time in sanctuaries dedicated to Aphrodite, Artemis, and Athena. Sometimes but not always pledged to celibacy, priestesses might be widows or in some cases married. Hardly cloistered, they were charged with managing orchards, vineyards, and estates associated with temples. A place was reserved for them in festival processions. Ceremonial seats were kept for them in the front rows of theaters. Priestesses from wealthy families spent large sums of money to build, maintain, and repair sanctuaries. In these and many other ways, priestesses transmitted beliefs, values, laws, and rituals that maintained, enhanced, and perpetuated the sacred "harmonia" that allowed communities to flourish.[9] We do not know what were the sacrifices, prayers, hygienic measures recommended by Diotima. Whatever they were, for ten years after her visit the city escaped contagion and was able to fend off attack from Sparta.

What might have prompted a young man of 29 to seek spiritual advice from a woman "honored by the gods?" In the *Phaedo*, waiting death by execution, Socrates recalled some of his youthful worries and concerns. When he was young, he told the friends gathered around him, he had been troubled by the prospect of mortality. He searched for an explanation of the generation and destruction that seemed so irrevocably woven into the fabric of natural existence. At first he looked to theories in natural science, searching for answers to questions like "With what part of the body do we think? Where does knowledge come from? What makes a human grow? What happens when we die?" A variety of answers were available. The philosopher Thales taught that water was the primal stuff of the universe. Anaximenes argued that air was the cosmic arche out of which all matter evolves. Anaximader made the primal material of the universe an indeterminate stuff that separates into various physical elements. Yet another approach was taken by atomists like Democritus and Leucippes. Matter, they said, was made up of tiny irreducible particles that interact, attach, and separate to each other to make physical elements and objects.

For a time, Socrates told his friends, he studied theories of the workings of the physical body, intricate movements of blood and air, mechanisms of the brain and heart. So deeply immersed was he that he all but lost touch with everyday sense experience. At the same time, the less he was satisfied with any of the answers on offer. In the end, he could no longer understand change at all—how it could be possible that one thing could divide into two or that hot could become cold. Either the world was one homogeneous static mass or a seething confusion of phenomena. In a fog of disappointed skepticism, he had one last hope. Someone mentioned a book by a philosopher who took a different tack, arguing that the cause of all things was a cosmic Nous or mind. Perhaps here, he thought, was an answer, a ruling intelligence that could make sense of the elements and mechanisms of human passion and human mortality and that could order love and death for the best. But again came disappointment. Like other natural philosophers, Anaxagoras went on to give the usual physical explanations of eclipses, meteors, sunlight, rainbows, and human physiology in terms of the air, water, and elemental particles (*Phaedo* 96–100).

In addition to ultimate questions of meaning and origin, might more personal worries have prompted Socrates to visit a religious authority wise in matters of love? Given the segregation of sexes in upper-class Athens, he would have had little opportunity to mix with females of his own age. At the time of the lessons, he was close to the age when he would be expected to marry. A wife would soon be found for him, and he would eventually sire three sons, but both before and after marriage he associated with a coterie of young aristocrats whose erotic impulses focused not on wives or female lovers but on adolescent boys. For a youth reaching puberty in Socrates's circle, pursuit by a well-connected older man and eventual acceptance of that man's favor constituted initiation into a segment of masculine society that looked to the military ruling caste of Sparta for many of its customs and values. Between the age of the onset of puberty at about 12 and the beginnings of a true beard, attractive boys were coveted "beauties" courted by older men who were potential lovers. They were "love objects" who, after suitable gifts and wooing, might be persuaded to accept a suitor out of "friendship" or φιλια. Short, squat, and pug-nosed, with a build better suited to military service than sexual dalliance, it is doubtful whether Socrates had been much pursued for his physical beauty, but by the time of his meetings with Diotima, he was proving to have an intellectual charm as or more seductive than a well-formed body.

Boy-love presented problems in love for both lovers and loved ones, some of which are touched on by various speakers in the *Symposium*. Does a boy who "gives in" and allows himself to be sodomized by an older man jeopardize his manhood and his reputation? Should all such liaisons be broken off once a youth grows a true beard? Should pederastic relations be limited to a few embraces and words of advice? If, on the other hand, a man preferred to spend time with and take advice from a woman as Pericles did with his companion and lover Aspasia, would that man be condemned for uxorious weakness? They were questions not unrelated to the more metaphysical worries recalled by Socrates in the *Phaedo*. What is this natural force that drives lovers together and perpetuates generation in the natural world? Is there no more to erotic attraction than crude physiological mechanisms? Is there any meaning to the endless generation, evolution, and destruction of living things? Is the attraction that draws us to another person only animal desire for pleasurable relief of tension in the genitals? Perhaps it is not so improbable that an Athenian man, hovering between youthful pedophilia and marriage, might feel the need for lessons in love, no surprise that he might consult religious authority or that two decades later, when called upon to stand up and give a speech in praise of Eros at a drinking party, he was able to relate what he learned from her almost verbatim.

PART I

Lessons in Love

CHAPTER 1

Daemonic Eros

Daemonic spirits are multiple and take many forms, and one of them is Eros.

—Symposium *202e7–203a9*

In fifth-century Athens, a symposium—a "συμ-πόσιον" or "drinking together"—was not the sedate affair it is today with professors sitting around a table discussing matters of scholarly interest. It was a festive even rowdy gathering not unlike a present-day fraternity party. Music played to spur on the drinking. Depending on the taste of the guests, handsome adolescent boys or female prostitutes and courtesans were brought in for entertainment. Poetry readings, speeches, and jokes went on long into the night. On the night in question festivities were somewhat subdued. Agathon was host and center of attention due to his win at the festival of Dionysius the night before, but both he and his guests were under the weather from a previous night's celebration. This was to be a smaller gathering of old friends and lovers. Pausanias, Agathon's long-time lover, was there. Eryximachus, a medical doctor, came with his young companion Phaedrus. Aristophanes, the comic playwright, attended, forgiven it would seem for his lampoon of Socrates and his followers in the *Clouds*. Socrates brought Aristodemos in from the street. Short, barefoot, and ill-bred, he was one of Socrates's "principle lovers at the time" according to Apollodorus, although what that meant in the case of Socrates was never quite clear. Another of Socrates's admirers, the rash young general Alcibiades, came at the end of the evening, stumbling drunk. All the men were in their thirties except for Socrates, who at fifty-some was by far the oldest.

According to custom, first came dinner, almost missed by Socrates who fell into one of his trances on a neighbor's porch and could not be persuaded to come in until midway through the meal. When everyone had finished eating, the dining tables were removed, and the first of the wine was brought out to be mixed with water in the great krater or wine bowl on the center table. A libation was poured, a prayer chanted to the gods, and the guests settled down to drink, tell stories, flirt, and, perhaps, fondle. Pausanias acted as "Symposiarque" or "King" of the party responsible for mixing the wine. He posed the question: What should be the manner of the night's drinking? In view of the previous night's excesses, should the strength and amount of wine be reduced? Eryximachus gave his expert medical opinion. Excessive drinking disturbs the humors and is bad for health. Given that they were all suffering from the night before, a change in program was in order. Send away the flute girl, said Eryximachus. Mix only enough wine to keep talk flowing. As to the subject of discussion, how about taking up a provocative challenge once laid down by young Phaedrus? Why had so many hymns of praise been written to other gods and so few to the great god Eros? Why not each of them stand up and make a speech in praise of love? Phaedrus as "father of the event" could go first. There was general agreement. They would drink moderately, and what could be more appropriate to celebrate the victory of Agathon—the "beautiful" "kalos" hero of the hour with his soft poetic way with words, slim figure, and short-cut beard—than a competition in praise of Eros? Agathon dismissed the flute girl. She could go pipe for herself, he told her, or, if she preferred, play for the women in the back of the house.

Phaedrus went first, taking a high moral tone. Erotic love of a man for a boy, he said, is to be encouraged because it promotes virtue. Always a man will want to look good before his youthful love object, and in the same way the courted youth will want to look good before his lover. This mutual admiration is especially useful in war as lovers try to impress each other with brave deeds and are even willing to die for their favored companion as the legendary Patrocles did for his intimate friend Achilles. Next came Pausanias, Agathon's older lover, introducing a note of moral caution. Not all Eros is good. In fact, said Pausanias, there are two Eros's just as there are two Aphrodites. A vulgar heterosexual Aphrodite "of the people" is opportunistic and of the body. Another heavenly Aphrodite Urania presides over erotic relations between virtuous men.

> Every one understands that there is no Eros without Aphrodite, and if that Being is not one how is Eros not two? There is the older Aphrodite born of no mother, but rather of Uranus, her we call Heavenly, and the younger Aphrodite born heterogenically of Zeus and Dione, her we

> call "of the people." It is therefore necessary that one Eros be called "of the people" as co-worker with one of these Aphrodites, and the other "Heavenly." (180d4–e1)[1]

As long as a man is inspired by the heavenly Eros, Pausanias went on, it is permissible for him to "have his way" with the boy he favors, since he does so not just for his own gratification but also to instruct the boy in manly virtue. On his part, a boy is virtuous in giving in to such a lover, as long as the age difference is not too great and the boy does it for self-improvement rather than for money or status.

The evening wore on. Eryimachus gave a long speech amplifying the distinction between healthy and unhealthy sex. Aristophanes livened things up with a humorous folktale in which eight-limbed hermaphrodites cut apart by the gods are made to run around forever looking for their "other half." After some jesting as to whether men might be cut in half yet again to produce a second race of male lovers, Agathon took center stage and spoke at length with poetic eloquence. Love is delicate, soft, and flowery; it brings peace and tranquility. Eros is the best of the gods and everything good and beautiful.

Socrates responded to the enthusiastic applause at the end of Agathon's speech with characteristic irony. Could the company think that he could have much to say in the face of such a "flood of eloquence?" Of course, he had the naïve idea that in explaining the nature of love one would state facts, not "run though standard attributes of power and virtue when the whole thing might be a pack of lies for all anybody cares" (198e). Nevertheless, he said, he was willing to take his turn, that is, as long as it could be on his own terms. First, he would have to determine what they were talking about, and to do this he would have to ask Agathon some questions to be sure they meant the same thing.

At this, knowing glances must have passed around the couches. Here he was, the Socrates they all knew and loved, not beautiful or eloquent like Agathon, not young or handsome like Pausanias, but with a wit so seductive that even at 50 young men still followed him around and hung on his every word. "Is the nature of Love to be of someone or of no one?" "Does not Love long for whatever it is love is of?" "And isn't the object of love what is beautiful and good, meaning that Love longs for what it doesn't have?" "But then how can love be either beautiful or good?" Stung by the paralyzing but thrilling stingray effect of the Socratic elenchus, Agathon good-naturedly succumbed. Confessing that he must not have known what he was talking about, he proclaimed himself defeated.

But now came the real surprise. Instead of proceeding on with more questioning, exposing ever more contradictory beliefs about love, Socrates took a new tack. He would leave Agathon in peace, he said, and pass on teaching he once had from a Mantinean woman "wise in matters of love and in much else" (201d4–5).

> I also had my lessons in love, so I now want to follow up the points on which Agathon and I just agreed by recounting to you on my own account, as well as I am able, the teaching she gave to me. First Agathon, as in your account, I will lay out what Eros is, what sort of thing, and then what is the work of Eros. The best way, I think, will be to give my account in the question and answer form she used then. For I spoke to her in much the same way Agathon just spoke to me, saying that Eros was a great God and everything beautiful and good, and she refuted me with the same arguments I just used against our young friend here, showing that, by my own account, it was neither beautiful nor good. (201e6–9)

And Socrates proceeded to recreate for his fellow guests the instruction he once received from a woman "wise in matters of love."

* * *

As Socrates tells it, Diotima's insistence that Eros is neither beautiful nor good took him by surprise.

> "What can this mean, Diotima?" I exclaimed, "Is Eros then ugly and bad?
>
> "Come now," retorted Diotima. "You do not speak well. Do you actually think that whatever is not beautiful must be ugly?"
>
> "Of course, I do."
>
> "And do you think that what is not wise is ignorant? Haven't you ever observed that there is something midway between wisdom and ignorance?"
>
> "Like what?"
>
> "You know, certainly, that to have correct opinion, if you can give no reason for it, is neither full knowledge—because how can an unreasoned thing be knowledge?—nor yet ignorance; for what just hits on the truth cannot be ignorance. So correct opinion, I take it, is in just that position, between understanding and ignorance."
>
> "True," I said.
>
> "Then don't compel what is not beautiful to be ugly," she said, "or what is not good to be bad. Likewise with Eros, when you find yourself admitting that it is not good or beautiful, don't then suppose it has to be ugly and bad rather than between the two." (201e10–202b6)

Socrates's puzzlement is understandable. Wasn't it a contradiction to say that what is "not beautiful" is not ugly, since "ugliness" is by definition "not beautiful." And didn't contradiction contravene the founding principle of philosophical rationality that "what is" cannot be "what is not?" A man is wise or not wise, that is to say ignorant. He cannot be both ignorant—not wise—and at the same time wise. It was the principle laid down by Parmenides, the philosopher from Elea whom Socrates had been so eager to meet ten years before, the philosopher who, Socrates recalled in the *Theaetetus*, made a profound impression on him.

> A feeling of respect keeps me from treating in an unworthy spirit Melissus and other followers of Parmenides who say that the universe is one and at rest. And there is one being whom I respect above all others. Parmenides himself is in my eyes, as Homer says, "a holy and awe-inspiring figure." I met him when I was very young and he was old, and I thought there was a kind of depth in him that was very noble. (*Theaetetus* 183e–184a3)

In his philosophical *Poem*, Parmenides condemned the wandering illogical thought of irrational masses who think ambiguously in terms of betweenness. He described a majestic goddess of Truth who sent down metaphorical horses and a golden chariot to bring the rational man up to heavenly heights on a "highway" that bypassed the confused and contradictory beliefs of ordinary men and women. As Parmenides put it,

> The horses that carry me as far as my heart can reach sped me on
> When they brought me to and set me out on the road of the spirit
> That bears the knowing man past all the towns. (*Frg.* 288)[2]

The goddess who greeted Parmenides at the end of his ascent did not appear like Aphrodite in an epiphany of golden wings and breath of fragrant incense. She waited behind a forbidding pair of massive bronze-studded gates. In that barricaded realm, far away from "the towns of men," she revealed to him the perfectly rounded and unmoving heart of "what is" and taught him to reject the wandering confused "mortal opinions" of common folk.

> [They] seeing nothing, wander of two minds
> for confusion in their hearts guides their wandering thoughts,
> And they are carried deaf as well as blind
> Astonished undiscriminating masses. (293)

Regardless of deceptive sensory appearances, from the standpoint of the new philosophical rationality, there could be no mingling or mixing between "what is" and "what is not." To Diotima's question—"Have you never noticed what is 'between' mortal and divine?"—Parmenides's goddess mandated an emphatic no. What is "between" cannot be real. A thing is or is not. To say that "what is" can also be "what is not"—that a man can be both ignorant and knowing, or eros both beautiful and ugly—is to be hopelessly confused. A thing cannot become what it is not. Whatever it is, it must have been before; it cannot be born, or change, or come to be something other than what it is.

One of Parmenides's targets was the theorizing of natural philosophers who thought they could explain the origin and nature of the cosmos with reference to the dynamic division or differentiation of one or more basic substances. Such naturalistic accounts of generation and change required the illogical thought that opposites mingle and mix, suggesting that "what is" could become "what is not." As Parmenides's goddess put it:

> [What is] never was nor will be, since it is now all united, one, and continuous. What birth would you want for it? How or when did it grow? I shall not let you think from "what is not," for "it is not" is not to be spoken or thought, and what need could make it grow, beginning from nothing, sooner rather than later. (296)

Whatever the senses might tell us, by the laws of Parmenidean logic, "what is" is one, indivisible, and unchangeable. The lesson was clear. True thought had to pull free from contradictory appearances; it had to focus on "One" knowable self-consistent and complete truth without wandering or wondering, unfulfilled longings, or giving birth to new and strange creations. The rational man had to reject wavering "betweeness." He had to learn to "look at things far off and securely present to the mind." He had to refuse to "cut off 'what is' from 'what is' by scattering diversity all over the cosmos and then drawing it back together into illicit unions" (313). Most of all, he had to refuse being "sent" by any daemonic spirit of love to mate male with female.

> In the middle of all of these is the Spirit (δαιμων) who steers all things originating loathsome birth and everywhere the mingling of things, sending female to mate with male and conversely males with females. (306)

As Plato tells it in the *Parmenides*, even at the young age of 20, Socrates had been quick to see some of the problems with this view. Did not Parmenidean logic threaten all human-to-human discourse, all conversation, all philosophical disputation? Yes, one might agree that appearances are deceiving, and certain knowledge of changing phenomena impossible, but how then could you define or distinguish justice, beauty, or goodness? How is it possible to make distinctions and differences that allow recognition of a friend or identification of an enemy, or mark out a good life as opposed to a bad life? And even if it were possible, how could the one true idea of justice be related to a just man? Are ideas thought phantoms or ephemeral shifting patterns only momentarily discernible in nature? Can anything true be said except "what is" is? Pressed to respond to Socrates's queries, Parmenides's prescription had been a course of mind-bending training in deduction, logical exercises in refutation that would prepare a philosopher to eventually grasp the truth. As skillfully demonstrated by Parmenides's companion and follower Zeno: spin out deductions on both sides of a question, practice drawing inferences from a statement and then turn around and do the same with the opposing statement, acquire masterful dexterity in manipulating abstract categories and generating contradiction.

Judging from the conversations remembered in the dialogues, on occasion Socrates took that advice to heart, spinning out refutation after refutation. Is true belief knowledge? Can a jury be said to "know" if they come to a right conclusion, but can give no "account" of that conclusion? Is knowledge possible? Once an affirmative answer proved contradictory, the negative could be subjected to another string of deductions also resulting in contradiction. Supporting these exercises were oppositionally defined ideas, perceived, as Socrates put it in the *Theaetetus*, not by the confused senses but independently by the mind "reflecting in itself on past and future." Exists/not exists, like/unlike, unity/plurality, even/odd, perceivable qualities like hard and soft as ideas in the mind might remain clear and distinct with no unsavory ambiguous mixing and mingling (*Theaetetus* 186b).

To evoke the transcendent reality of this conceptual scaffolding, Socrates would often resort to "dreams" and "visions." In the *Theaetetus*, he imagined first concepts or ideas not "knowable" in the sense of definable in terms of other concepts, but directly perceived by the mind and used to give an "account" or argument as to why a thing is just or unjust, right or wrong. In the *Phaedrus*, he produced a mythical "word painting" of winged souls contemplating eternal icons of Justice, Knowledge, and Goodness. In the *Meno*, he projected the possibility that belief might be "tethered to" knowledge gained in prophetic vision or by recollection of an eternal Form from previous existences (97d). In the *Cratylus*, he envisioned a rational language of the mind as an alternative to the changing meanings of natural language. In the *Republic*, he projected an ideal ruling caste, educated in rigorous courses of mathematics and dialectic so as to achieve a final vision of the Form of the Good that would allow them to "run the gauntlet of all tests and strive to examine everything by essential reality and not by opinion, holding the way without tripping up in reasoning" (533e–534).

But now, not only had his mentor in "matters of love" placed Eros back in the ambiguous unknowable murky "betweenness" not to be spoken of by the true philosopher, she questioned what seemed generally accepted truth.

> "But," I said, "doesn't everyone agree that Eros is a great god?"
> "You mean everyone who doesn't see? Or are you including those who do see?"
> "I mean everyone together."
> "But Socrates, how can those agree that he is a great god who think heis no god at all?"
> "Who are they?"
> "Why you for one, and me for another."
> "What is it you are saying?" (202b6–c5)

And it was at this point that Diotima took Socrates step by step through the argument he would later use on Agathon, proving herself, as Socrates remarked, as good as any Sophist at refutation: All gods are happy and beautiful. Those who are happy must possess what is good and beautiful and can want for nothing.[3] Socrates had agreed that Eros desires what is good and beautiful. Therefore whatever he thought he believed, he was not someone who actually did believe that Eros is a god.

Socrates's assumption that Eros was a god was based on more than faulty logic. Starting early in the fifth century, artists had begun decorating vessels made for use at symposia like Agathon's with idealized semidivine depictions of the slim, naked "kalos" youths admired and courted at such gatherings. Wings sprouted on the backs of boys, indicating the flight and pursuit of pederastic courtship. On one cup, two young male lovers, one winged, one not, copulate. On another, a naked winged youth represents the transcendent spirit of desire hovering over male lovers engaged in sexual intercourse. On yet another, plural "*erotes*" in the guise of two or more winged youths carry flutes and lyres to make the music that stirs on sexual desire.[4] The deified Eros was the idolized beautiful "*kalos*" adolescent male, admired as erotic object by the circle of Athenian men frequented by Socrates. It was the "Eros, your and my master, observer of beautiful boys" to which Socrates dedicated his mythic hymn of love in the *Phaedrus*. Agathon's speech in praise of Eros at the symposium could be taken as praise of his own image as still the "youngest-looking," "the prettiest, "the most tender" of coveted boys.

But, as Diotima notes, "everyone" in Socrates's circle was not every one in Athens, or everyone in Greece. Eros was seldom mentioned by Homer and never as a deity. In Hesiod's *Theogony*, Eros is not a god but a primal cosmic force of attraction present at the genesis of the material world. Nor was Eros worshiped as an independent deity in any cult or religious observance.[5] As the chorus comments in Euripides's *Hippolytus*, "sacrificed oxen might pile up in sacrifice to Zeus in his temple by the Alpheus River and to Apollo at Delphi, but no devotions are paid to Eros with its power over men" (538ff). Eros's pairing with Aphrodite on inscriptions in no way established Eros as anything more than an attribute, emanation, or attendant of Aphrodite. Eros as a deity was the invention of a group whose ideas and practices did not reflect those of the general populace.[6]

The god-like status of Eros in the person of an attractive adolescent boy was compromised by more than illogic and more than lack of official cult. Greek cities enacted statutes banning older men who were not teachers or trainers from schools and gymnasiums where young boys were educated and trained. As described in Plato's *Lysis*, parents hired "guardians" to see boys safely through the streets and protect them from molestation. Only the occurrence of a festival day allowed Socrates and his friends to gain entrance to the gymnasium to watch and admire Lysis and his young friends exercise naked. Admirers of young boys themselves acknowledged a need for ethical restraint. Taking advantage of boys younger than 12 was frowned on. Gifts were allowed, but no money should pass hands, especially in Athens where conviction for homosexual prostitution meant a loss of rights of citizenship. Couples like Pausanias and Agathon, Zeno and Parmenides, and Eryximachus and Phaedrus maintained intimate relationships into middle age, but in most cases, pederastic coupling was supposed to end when a boy reached manhood. At the opening of the *Protagoras*, Socrates is chided for his pursuit of Alcibiades even as Alcibiades is growing a beard and is forced to defend himself by quoting Homer. "The most charming age of a youth is when he is growing his first beard" (309a). How far pederastic intimacy should go was a matter of dispute even among its practitioners. Is any degree of sexual contact allowed as long as the boy's welfare is taken into account? Or should intimacy end with affectionate kissing and fondling as Socrates proclaims in the *Phaedrus* and the *Republic*? Is orgasm permissible as long as there is no penetration of a boy's body parts? Should the alternative of intercourse "between the thighs" depicted on some vases and kraters be substituted so as to leave a boy's orifices intact? But for many citizens of Athens, talk of ethical restraint was nothing but philosophical hypocrisy, the clothing of problematic objects of sexual desire in the trappings of divinity.

Socrates's question remained unanswered. If Eros is not an immortal god and not mortal either, if Eros is not all-beautiful but not yet ugly, what then is it? Having cleared the air of false conceptions, Diotima was now ready to give him a definitive answer. "Eros is a Δαίμων μέγας, a great spirit, for the whole of the spiritual is between divine and mortal."

* * *

Only later in a Christian era would "daemon" become "demon," the personification of evil in the most virulent of forms. The verb from which the Greek noun comes is experiential, indicating moments of divine revelation that bring a sense of being in touch with influences outside ordinary experience. The adjective, δαιμόνιος, characterizes what is heaven-sent or miraculous, whether beneficent or admonitory. For Hesiod, the spirits of men of a lost Golden Age were daemons and. as such, possible channels of communication between humans and gods. In Homer, the daemonic is a kind of divine frenzy that overcomes warriors in the heat of battle. In Aeschylus, the Eumenides or Furies are daemonic spirits that haunt individuals and families who commit grave crimes of murder. Socrates claimed daemonic visitation in the form of an inner voice that warned him when he was about to do something wrong. The word "eudaimonia," translated as human well-being or happiness, combines "eu," or favorable, and "daemon," to indicate a state of being spiritually blessed. "*Deisdaimonia*," dread of the daemonic, is superstition as opposed to piety.

The key to understanding the daemonic is the principle, asserted by Diotima, that although earthly creatures directly mix and mingle, gods and humans do not.

> God does not mingle with humans, but through the daemonic comes all associations and discourse between gods and humans whether

> asleep or awake. Whoever is wise in these matters is a spiritual man, whereas knowing technologies and how to make things is for the merely mechanical. (203a8–9)

Gods do not meet with humans the way a person meets with another person or thing. Humans do not visit heaven to commune with gods. In the fanciful stories of Greek poets, gods "mingled" all the time. They disguised themselves as humans to advance the causes of favorites. They struck down humans they did not like. They got angry and sent floods and wildfires. They raped mortal women and seduced mortal men. They punished those who neglected their worship and roared thunder from the sky in anger. Diotima's daemonic comes not from storytelling but from priestly practice. The naked young pubescent Apollo's so admired and pursued by Socrates's circle were not gods, any more than were the chubby toddler Cupids later depicted on the bedroom walls of Roman villas. Nor do gods live on mountaintops to which a human might conceivably ascend or from which a god might come down. Divine and human are realms or dimensions of reality that in no way come face to face. Knowledge of the daemonic requires a spiritual wisdom that is unlike craft or technical skill that involves the manipulation of material objects.

It is not that gods happen to inhabit a place too far away for humans to see or reach. Gods are not, as Epicurus sometimes suggested, an alien immortal species of beings living on some distant planet with no interest in human life. If this were the case, there would be no reason or basis for any form of religious practice or worship. Humans would live for survival; measuring out pains and pleasures until death put an end to all sensation.[7] Eros would be no more than a "mortal" physiological urge or θύμος rising up in animal bodies. Humans might develop rationality and foresight, form cooperative social networks to aid survival, invent tools and mechanisms to help them survive. With language, they might come to imagine and tell stories of superhuman beings who will reward them with immortality and transport them to other worlds. But neither a materialist understanding of humans as evolved animals nor a belief in remote supernatural beings can accommodate the fact of spirituality as explained by Diotima.

Socrates's questions in response to Diotima's identification of eros as daemonic—"Having what power?" "What father does it have? What mother?"—suggest that he still thinks in mythic terms. These were questions asked of humans and deities modeled after humans. A man's parentage identified his place in a social network of relation and patronage. A man's "powers" in athletics, war, or speech-making further differentiated him as an individual. Poets invented such parentages and powers for the gods in their stories. In the *Theogony*, Hesiod laid out imagined family histories, adapting and aggrandizing human sexual behavior in ancestor stories of rape, entrapment, adultery, and castration. Zeus was born of Cronos and Rhea and ended by killing his father with a thunderbolt. Aphrodite stepped, motherless, from the sperm foam of Uranus's severed penis cast into the sea. Athena emerged from the forehead of Zeus after he swallowed and ate her pregnant mother Metis. All involved plots not so different from tales of cannibalism in the house of Atreus or deceit in the royal family at Thebes.

Stories require plots and plots require characters, and poets duly constructed personalities for their divine actors. "Powers" were meted out, powers greater in extent than human powers but not different in kind. Zeus as warlord was given the weapon of the thunderbolt; Poseidon, god of the sea, could blow and cause floods and storms. Athena could disguise herself as a Trojan prince. Aphrodite could seduce mortal men. Ares was bellicose. But these were narratives invented and performed by storytellers. Written down, they could be used to entice schoolboys to learn to read. Organized into family histories by Hesiod and Homer, they symbolized the paternal authority of Dorian Zeus over local deities like Hera or Rhea. Sophisticated listeners and readers might find in them some allegorical truth. Intellectuals might claim to discover traces of protoscience or history. It was speculation that in the *Phaedrus* Socrates himself dismissed as a waste of time (230a).

Divinity as understood by Diotima has no parentage and no powers in any human sense. It is not one, or three, or any number of divine persons, nor is it definable in human or material terms. At the same time, it is connected with the human world by channels of daemonic spiritual communication, including practices traditionally defined as religious. Responding to Socrates's question about the "powers" of the daemonic, Diotima answered:

> Interpreting and carrying human things to the gods and divine things to humans—prayers and sacrifices from humans, rules and forgiveness from the gods. Midway between human and divine, the daemonic allows each to fulfill the other so that they are bound together. By way of it comes divination, priestly arts such as sacrifice, ritual, and incantation, and also prophecy and magic. (202e6–203a8)

Festival celebrations, prayer, sacrifice, divination, ritual, cult, prophecy, contemplation of sacred objects, all are "between" gods and humans with the function of bringing human and divine into alignment, "binding them together," as Diotima puts it, so that each "fulfills" the other.

Diotima's identification of religious practices as daemonic reflects patterns of popular Greek piety. Worshipers came to shrines like Aphrodite's cave sanctuary below the Parthenon or to her temple outside the city walls not to placate a divine ruler but in solitary vigil. Meditation in sacred places allowed the mind to settle, contemplate, work through irrational fears and desires. Prayer bound aspiration and method together so that a worshiper tapped reservoirs of resolve and insight, grasped patterns and symbols that helped make sense of human experience. At seasonal festivals, hymns sung or chanted with neighbors bought communities together, expressing shared values and inspiring creative endeavors. Sacrifice was celebration of a communal sacred meal, sign of respect for the life of an animal and an acting out of ties between divine and human. Incantation, processions, sacred dance, choral odes mimed the divine beauty of order and symmetry. Prophecy was the utterance of inspired human beings who at mantic moments grasped hidden possibilities, dangers, patterns, symbolic meanings. A person might feel a special affinity for one set of images, practices, or religious figures. Particular religious observances might be native to a particular geographical location or national history. As daemonic, such preferences did not preclude devotion to other gods in other places.

But although prayer, cult, and prophecy are identified by Diotima as important avenues of daemonic spirituality, it is eros she identifies as the "great spirit," the Δαίμων μέγας. To explain its source in human life, she invents a parable. At a festival of Aphrodite, a guest "Resource" or "*Poros*" drinks too much and falls asleep in the "garden." "Need" or *Penia*, always begging and poor, finds him there and mates with him. From that mating comes eros. Humans are needy and lacking, but they are also clever, able to find resources that give rise to strategy, practical foresight, and ways to bring about what is lacking. When human need and human resource come together, the result is eros, which is neither abject despair nor unfocused misguided plotting. If "*penia*" is the sense of something lacking, resource rescues need from passive complaint and puts it into creative motion.

Diotima's "*penia*" and "*poros*" are not superhuman personages, but existential aspects of the human condition that come together to create the phenomenon of eros.[8] The metaphorical lying down of Need at the side of Resource and the resulting "birth" of eros is not an account of sexual adventure. Nor does Diotima's answer to Socrates's question about the "powers" of eros give a list of traits, like Agathon's "everything beautiful, gentle, and good." Need and lack move living things to survive long enough to reproduce; mated with resource, they move human beings to aspire further to social, cultural, artistic achievement beyond the requirements of survival and reproduction. In Diotima's teaching, there is no final supernatural Providence, no apotheosis of desire. Inevitably, in the process of natural existence and in response to what could be but is not, disappointment and frustration return and new infusions of resource must be found. Eros ebbs and flows according to the rhythms of natural life, at one time lonely and bereft, at another swelling with hope and energy, but always in movement and never static. As Diotima puts it:

> From birth neither immortal nor mortal, eros in the same day is flourishing and alive when he is well off, then dying and again reviving with the force of his father's nature. The resourcefulness he has eternally ebbs away, so that at no time is eros ever either without resources or rich and, in the same way, always between wisdom and ignorance. (203e1–7)

In Diotima's parable, divinity in the form of Aphrodite plays an essential role. *Penia* and *poros* come together under the goddess's auspices, at a celebration of her "coming" reminiscent of the great annual festival of the Aphrodisia at Aphrodite's shrine on the southwest slope of the Athenian Acropolis.[9] On the day of the festival, hymns of praise and joy greeted the statue of Aphrodite as it "came out" of the inner sanctuary to be carried in procession through the streets. Wine, music, and dancing enhanced the celebratory mood. Love affairs blossomed. Those lacking or needy found the resource to achieve what they desired. Cleverness, which alone might dissipate into thoughtless activities, overcome with the intoxication of beauty, was waylaid by need, and eros came into being as directed and focused desire.

Without need, eros would not exist. Gods do not love because they lack nothing. Ignorant mortals who do not feel in need do not desire knowledge, which is what, remarked Diotima, makes ignorance so hard to deal with.

> It is this way: No god pursues wisdom, or desires to become wise. They are wise, nor would anyone wise be a lover of wisdom. But neither do the ignorant love wisdom or desire to be wise, and it is this that makes ignorance so troubling that a person neither beautiful nor good, nor thoughtful are satisfied with themselves and do not desire what they lack. (203e7–204a7)

But need alone is powerless. It can only metaphorically sleep on doorsteps and beg for favors. Paired with cleverness and resource, need becomes brave and active, venturing out to achieve what it desires whether wealth, honor, love or wisdom.

Needy, searching, waylaying: it was hardly the heroic image of the rational philosopher projected by Parmenides, but Socrates's protest at this point—"What then are philosophers if they are neither wise nor ignorant?"—gets a rebuke from his mentor.

> "Come now," she answered, "a child ought to be able to see it by now. They are between, just as eros is between. Wisdom is of the most beautiful things, and eros is directed toward the beautiful, so that by necessity philosophy is erotic and as such between wisdom and ignorance." (204b1–8)

In place of the Parmenidean philosopher directly acquainted with Truth, Diotima's "lover of knowledge" is neither ignorant nor all-knowing, but always between, needy but also clever, often disappointed and poor of spirit, but always finding the confidence to think again. Diotima diagnoses the mistake made by Socrates and his friends.

> That you would have formed this other idea of Eros is not surprising. You supposed, judging from what you have said, that the object of love was love and not the loving, This made you, I imagine, claim that eros is all-beautiful, What is loveable indeed is beautiful, elegant, perfect, and most fortunate, but loving is different as I have shown. (204c1–7)

Bedazzled by their beautiful boys, making them into semideities, Socrates and his friends confused the beauty of a loved one with loving itself.

Socrates's next question—"Well then, Mantinean, you speak well, but if Eros is as you say, what use is it to us humans?"—comes as a matter of course. If eros is not a great god with the power to reveal the truth, not a deity one can pray to or ingratiate with offerings, but simply loving with all its bodily and psychic yearning and struggling, what good is it? Why yearn for the latest beautiful boy? Why want so passionately to know the true nature of justice? Why not do away with eros and concentrate instead on useful "technologies and manufactures"? They were questions that the pedophilic lovers of Socrates's circle might well ask. For them, eros was a game of love played between men and boys. You court a boy. You give gifts. You mentor and tutor. You impress. You admire. If you are successful, the boy becomes yours, yours to enjoy, yours to mold in your image, yours, perhaps, as lifelong admiring friend and ally. But once that yearned-for moment of pleasure was over, what did the lover have? If the boy was constant, his lover had the expectation of future moments of pleasure. But had the effort been worth it? Might not even greater pleasure be possible with another boy more compliant or a better protégé? What in the end was the point of all the endless "cleverness" of eros, the enticement, flattery, rivalry, preening, the competing for possession of someone seen as beautiful?

For once Diotima approved the direction of Socrates's thought.

> "Yes," she replied, "this is indeed the next question to ask and the question on which I will now try to instruct you. Eros is this, and of this origin, and is of the beautiful, as you have agreed."
>
> "Now," she went on, "what if someone were to ask us this: Socrates and Diotima, how is Eros of beautiful things? Or to ask the question more clearly, what is it that a lover of what is beautiful wants?"
>
> "That they might become his," I answered.
>
> "But your answer only provokes a further question," responded Diotima. "What is it to this lover when what is beautiful becomes his?"
>
> "And that question," Socrates confessed to his audience at the symposium, "I declared myself at that moment completely unable to answer." (204d8–11)

* * *

At this point in the lessons, Diotima's teaching takes a turn that causes endless trouble for scholars. She changes the subject: "Come now, Socrates, I ask you. The lover of the good, what does he want?"

For a time Socrates's answers come without hesitation: "What does the lover of good things want?"—"That he can have good things." "And what will he have when he has them?"—"Even easier, he will be happy." "And do you think that this wishing and this loving is common to all humans?"—"Yes it is common to everyone."

They were standard answers. Everyone desires the good. What is lacking is the knowledge of what is good, really good not just what people think is good. It must have seemed to Socrates, as it has to many readers, that the conversation shifted onto solid ground.[10] No more strange talk of daemonic love and beauty. The good was now in question, the good man, the good life, subject of debate in so many conversations recreated in the dialogues. Up to now, Diotima had always paired good with beauty. Gods are beautiful and good. Lovers lack what is beautiful and good. Now with "good" singled out, familiar questions might be asked and answered. Do we always desire the good? Are some desires evil and bad? How can we know that what appears good really is good?

But instead of continuing on to distinguish the true good from mere opinion about the good, Diotima immediately returned to eros and a lesson in linguistics. Socrates had agreed that wanting the good is "common" to all humans. But what about eros? Why don't we talk about love in the same way?

> But, at the same time, when we say a man "loves" we do not mean that everyone loves the same things, but only that some love and some do not. We single out one form of eros and apply to it the name of the whole. And there are other words we misuse in the same way. (205b5–8)

To further illustrate the "misuse" of words, Diotima uses another example, the Greek word "*poesis*." The primary meaning of "*poesis*," she reminds Socrates, is general, a creative bringing into existence of what did not exist before. As such, the term is applicable to many different arts and creative making. When, however, one kind of "*poesis*"—verse with music and meter—is singled out as the only poetry, as it often is, the reference of "*poesis*" constricts and its general meaning is lost. One form of creative making is privileged and only those who create what is narrowly defined as "poetry" are called poets. Poetry's link with other arts is forgotten and the general nature of *poesis* is missed. The same, concluded Diotima, is true of "eros."

> Principally, eros is all desire for good things and happiness, "Eros most great and clever." But those who turn to it in different ways—money-making, athletics, philosophy—are not called loving or lovers. Instead only those who pursue love in one of its many forms are given the name of loving or being a lover. (205d)

It was a "misuse" of language with particular reference for the circle around Socrates. For them, "Eros" had come to mean the narrowly focused passion that drove men like Hippothales in the *Lysis* to make a fool of himself over a boy, neglecting business and family. But does everyone love like Hippothales? Even more important, does not this narrowing of reference obscure the true nature and function of eros? Does it not miss the daemonic spirituality of eros? The mistake is both in the use of a word and in practice. Taking "eros" to mean only a desire for sexual contact, or more narrowly for proprietary rights in one particular boy body, puts language out of joint with human aspiration, limiting eros to one obsessive pursuit and obscuring the continuity of sexual desire with desire that energizes many kinds of human relationship and activity.[11] Diotima prods her student: "We only value what is ours, I imagine, when we call what belongs to us good and what belongs to someone else bad. Or do you disagree?"

Again, as Socrates tells it, he had no recourse but to agree. "Humans love the good."—'"Yes". "And by necessity, they must love the good to be theirs."—"By necessity." "And they love it to be for them always."—"That too." "And they want it to be for them eternally."—"That is the truth." But if he thought the lessons were over, he was wrong. Diotima's next question again took him by surprise.

> "Now. If Eros is always for this," she went on, "what is the way of life of those who pursue it, the work (πρᾶξις) of all this eagerness and straining called Eros? What is it that eros aims at? Can you tell me?"
>
> "Hardly, Diotima. Would I be sitting here amazed at your wisdom and wanting to learn about these things, if I could."
>
> "I will tell you what is the work of love," said Diotima. "It is the bringing forth with the beautiful both in body and in spirit."
>
> "Prophetic powers I would have to have to understand what you are saying," I said. "I do not understand." (*Sym.* 206b1–12)

CHAPTER 2

The Work of Love

The nature of mortal beings is ever to try as much as possible to avoid death, and the only way to accomplish this is by generation in that it can always leave behind something new in place of the old.

—Symposium *207d1–3*

Socrates is not alone in failing to understand Diotima's teaching at this point. If Diotima's suggestion of continuity between sexual desire and desire for the good has perplexed scholars, even more disconcerting has been her abrupt abandonment of "the good" and return to beauty. Are "good" and "beauty" no more than different names for the same thing? Do the two categories overlap with some things both beautiful and good, and others beautiful but not good, or good but not beautiful? Is it a question of genus and species with the "good" the more general category and "beauty," along with virtue and justice, one of the concepts that fall under it? But even stranger than mixing beauty with good has seemed the introduction of generative language. "Bringing forth," "τόκος," a word referring both to the birth and the siring of children, both to the reproductive act and its resulting offspring, to desire and action? Was Plato confused, putting words into Socrates's mouth that make no sense?[1]

Incomprehension on the part of Socrates might be understandable. In upper-class Athens, the begetting and raising of young children went on far removed from the schools of rhetoric, counsel meetings, public venues, and battlegrounds frequented by men and boys. Socrates's midwife mother provided him with a few metaphors for the "delivery" of the predictably stillborn ideas of shallow thinkers, but like other men of his circle, he would play no role in birthing or early parenting. Begetting, for the men he knew, meant no more than periodic visits to the women's quarters for intercourse with one's wife. The act of sex that the men around him called by the euphemism "*Aphrodisiazing*" had nothing to do with reproduction. You felt a need, a bodily tension. You learned ways that tension could be relieved, perhaps alone, or with one of the prostitutes available in the streets, or with a younger boy.

And perhaps even that had been something of a mystery to him, all that hot pursuit of beautiful boys. In the dialogues, Socrates is often depicted as doing little more than play at flirtation: disputing with Alcibiades for which of them would sit next to Agathon at the symposium, admiring with Critias the body of the boy Charmides, drawing aside the young Lysis and his friend in the gymnasium, but always with a certain eyebrow-raised detachment. Certainly he had no lack of opportunities for sexual engagement, as was made clear by Alcibiades later at Agathon's.

> I'd been bitten by something more poisonous than a snake, in fact mine was the most painful bite there is. I'd been bitten in the heart by Socrates' philosophy which clings like an adder to any young and gifted mind it can get hold of and does whatever it wants with it. (218a)

As Alcibiades tells it, he offered himself to Socrates, invited him to dinner, lured him into bed, even though it was the older man who was supposed to do the pursuing. But it was all for nothing. Socrates lay all night beside him without making any move of "Aphrodisiazing," no stroking and arousing, no penetration of body parts, no passionate release. Alcibiades concluded:

> You know he doesn't care a rap about good looks—on the contrary he looks down on them—and on wealth and any of the honors most men care about. He doesn't give a damn for anything like that, or for any of us either—yes I'm telling you—he spends his whole life playing his little game of irony, and laughing up his sleeve at all the world. (*Sym.* 216e)

In the *Lysis* with Hippocratus, Socrates made a great show of distracting the boys, dazzling them with arguments that made a reasoned choice of a favored "friend" so complicated as to seem all but impossible. He claimed to be illustrating to the besotted Hippocratus a better method of seduction. In the end, it could have been no more than a harmless way of marking time and keeping the boys occupied until their guardians came to take them home. And that afternoon on the banks of the Ilissos in the *Phaedrus*, with young Phaedrus so full of praise for Lysias's clever seductive speech in favor of a suitor who makes no pretense of being in love? Was it simply to distract an impressionable young man that Socrates improvised rival speeches of his own on either side of the question? At the end, his greatest worry seemed to be that Phaedrus took his speech in defense of love seriously, even after he had pointed out to him the moral of the exercise: making speeches inspired by myths and metaphors can be as misleadingly seductive as clever arguments. Young men are mentioned in various of Plato's dialogues as Socrates's boy-love of the moment. In the *Phaedrus* it is Isocrates, in the *Symposium* Aristodemos, in the *Protagoras* Alcibiades. But if we can believe Alcibiades in the *Symposium*, Socrates was one "philosopher" who, finding himself in bed with a boy under cover of darkness, did restrict himself to the chaste embraces recommended in the *Phaedrus* and *Republic*.

As for birth, it was exclusively a woman's business, and Diotima would have to explain at some length the place of generation in human life.

> Let me make it clearer, she said. All humans are pregnant, both in body and in mind, and when we come to a certain age, our nature longs to give birth. But to give birth is impossible with someone ugly, it can only occur with someone beautiful. The coming together of a man and a woman is a begetting for both. It is a divine matter, this urgent desire for pregnancy and giving birth, an element of immortality in a living mortal being and it cannot happen with what is discordant. What is ugly is what is out of joint with divinity whereas the beautiful is what fits with divinity. As Moira and Eileithyia, Kalloni or Beauty is this bringing forth. (206c–d2)

Men as well as women "pregnant?" Not only the female desiring to give birth, but "human" "nature" desiring it, presumably the same in men as in women? A "begetting" that is a mutual endeavor not the physical insertion of sperm into a woman's womb? Συνουσία, coming together, meaning not only physical intercourse, but conversing together, creating things together? Coming together only possible with someone beautiful? Is this to say that a man cannot impregnate a wife he finds less than beautiful? If coming together is only to get a son, or enjoy a moment of pleasurable release, it would seem that any partner, any womb, any orifice should do, erotic attraction or not, although in the case of arranged marriage there might be a benefit in achieving in-laws with status and influence.

But neither acquiring an heir nor a moment of pleasure is the aim of erotic desire as explained by Diotima. Eros is not an urge like hunger to be relieved by a physical act, or part of the necessary body mechanics of acquiring an heir. It is a response to another living being seen as beautiful, someone with whom we "fit." The word she uses is αρμόττον, fitting together like a well-made joint in carpentry with a wide extension of meaning applied to betrothals and marriages, to the management of towns and provinces, harmony in music, even to the proper fit of a suit of amour. Its opposite, αναρμόστον, is what does not fit, is what is out of joint, poorly fitted, likely to come apart at the slightest strain. We pull back in the face of ugliness, depressed and unfulfilled, but we are drawn forward and energized in the presence of beauty, excited by the prospect of new experiences and successful joint endeavors.

Sacred images weave throughout Diotima's explanations. Moira, a term of ancient Minoan/Mycenaean origin, is fate and limit, the fabric of experience entangling a man or woman in a course of a life with an allotted end. Eileithyia is the miracle of childbirth and midwifery, associated with the delivery and emergence of the newly born child. Both are ruled by Kalloni, beauty that arouses erotic desire and shapes the course of a life. As longing for and expectation of what might be brought to life, beauty engenders Eileithyia, midwifery that facilitates successful delivery of the newborn. Beauty, as "First of the Fates," Aphrodite's title at her temple near the Ilissos river, inspires the coming together, sexual attraction, and mating of need and resource that makes generation possible.[2]

Beauty is not an idea to be contemplated but the powerful attractive force that puts lovers into motion, as opposed to ugliness that represses and dampens spirits.

> "When the pregnant one comes near to someone beautiful she becomes cheerful and flowing over with desire to beget and give forth, But when she comes near to someone ugly, she coils into herself angry and depressed, laboring under the burden of unborn young. A person, pregnant and swelling with new life trembles and flutters for the beautiful one, hoping to be set free from such a great burden. For you were wrong, Socrates, to say that eros is of the beautiful."
>
> "What is it then?"
>
> "It is the engendering and begetting with the beautiful."
>
> "So be it," I said.
>
> "Of course," she went on, "And why of engendering? Because engendering is what is forever becoming, what does not die in mortal life. We must by necessity desire what does not die as much as we desire what is good, since eros is for the good to be for us eternally. It necessarily follows then that love is of what does not die." (206d–207a4)

Eternity here is not the eternity of Platonic forms set in the hard stone of universal definition, but a good always coming to be, generated by human effort and striving. Diotima prods Socrates to look around and observe natural life.

> What do you think, Socrates, is the cause of this eros and this desire? Haven't you noticed the strange state animals get into when they desire to give birth, whether land animals or birds? How afflicted and erotically disposed they are, first to mate with another and have intercourse, and then to find nourishment for their newborn young, for which they are ready to fight fierce battles, even the weakest against the strong, ready even to give up their lives or to suffer starvation themselves, do anything, if they can only take care of their young. In humans, one could say that it is reasoning that pushes one forward, but in animals what can be the cause of this amorous state? Have you got an answer? How can you ever think to become skillful in matters of love if you cannot conceive of this? (207a7–c2)

Had he noticed? At the time of the lessons he had no wife or sons, although by the time he gave his speech at Agathon's symposium, he was married and a child was on the way. Even then, judging from the curt dismissal of his wife Xantippe from his deathbed as described in the *Phaedo*, he had little love or concern for her. Perhaps he passed by mothers begging for food in the streets or heard stories of famine in the countryside. As for animals, never in any of the conversations depicted by Plato did he take any notice of them. By his own account he was a city man with nothing to learn from trees, or new lambs frisking in spring meadows, or nesting birds as depicted in the cornice over the epistyle of Aphrodite on the Acropolis. He understood that sexual intercourse was necessary if a man was to sire a son. He noted the eagerness with which Hippocrates pursued the boy Lysis. He recognized the seductive come-on in the speech against love read out to him in the *Phaedrus.* Catching a glimpse of the boy Charmides's naked body under his garment, he was able to quip:

> I could no longer control himself. I thought how well Cydias understood eros, when in speaking of a beautiful boy, he warns someone not to bring the fawn in the sight of the lion to be devoured by him, for I felt that I had been overcome by a kind of wild beast appetite. (*Charmides* 155d–e)

But chided by Diotima, he was ready enough to confess his ignorance.

> Once more I replied that I did not know, and she replied, "How are you going to master matters of love if you have no idea of this?"

> "But it is for this, Diotima—like I said before—that I am coming to you, because I know I need instruction. So tell me the cause of this and other things involving matters of love." (207c3–9)

Ερoτικα, "matters of love." How hard it is to make sense of the name Diotima gives to her teaching when the meaning of "erotika" has narrowed to denote no more than the jaded images of pornography. Diotima's "erotica" encompasses all the generative energy that moves and empowers mortal life.

> It is only for a short time that a living being is called alive and itself, in the sense that a person is said to be himself from childhood until old age. Yet though he is called himself, he never is the same. He is continuously becoming someone new and losing things, such as hair, flesh, bones, and body, and notice, not only in body but in psyche as well. None of his manners, or habits, desires, pleasures, pains, or fears ever stay the same, but some grow and others perish. (207d4–e3)[3]

In Diotima's teaching, there is no escape from the generation and destruction that worried Socrates in his youth. Mortal beings not only live in a natural world in a state of constant change, they themselves constantly grow and age in body, emotions, and behavior.

And change is as true of minds as it is of bodies. There is no pure life of the mind where ideas stand in permanent oppositional order as was "dreamed" by Socrates in the *Cratylus*. As responding living beings, both what we think and what we know constantly changes, as do our bodies, feelings, and habits.

> And here is a stranger fact. Even with things that we know, not only do some of them grow and others die away, so that even in what we know we are never the same persons, but the same is true of every bit of knowledge. What we call studying implies that knowledge is leaving to make way for new knowledge, while studying substitutes a fresh knowledge in place of the old, so preserving knowledge to make it seem the same. (207e3–208a9)[4]

In Diotima's teaching, there are no Pythagorean dreams of an eternal mathematical order, no winged *erotes* following in the trains of gods, no guardians of the state traveling up from the dark cave of human illusion to a sunlit world of pure Reason, no eternal objects of desire that never age, decay, or disappoint. Her solution to the problem of generation and destruction remains rooted in reality. Transcendence and salvation are found not in dreams or fantasies but in the birth and destruction that Socrates found so unsettling as a young man. In injury or disease, the body regenerates. Cells divide and are reborn. New skin appears, wounds heal. Incapacitated by grief, the mind heals. A diseased limb is cast off; life is renewed. New thoughts replace regret at what is past. Memory substitutes for the lost presence of a loved one, leading to new determination and resolve. In the face of loss, living beings recover and move on to new projects and creations.

In response to Socrates's questions—"Is it so?" "Can this really be true?"—Diotima prompts him to look and observe what goes on around him. Look at the men you know, she tells him, how eager they are to make names for themselves. Look at heroic figures of the past. Why would they strive so to excel? Why would a man want to lie down and die in the dust like Achilles after the death of his comrade Patrocles? Why would a woman like Alcestis agree to die in the place of her husband? The answer is simple, we want the good, and we want it forever. Not being gods or able to "mingle" with gods, the only way to achieve this as mortals is to leave behind something of value that survives.

> Every mortal being is saved in this way not by keeping the same forever like a god, but by replacing what is gone or old with something fresh in its place, In this way, Socrates, a mortal being shares in immortality both in its body and in other respects and it cannot be done in any other way. So do not be surprised that everything loves its own offspring, since all are moved by this eagerness and love with a view to immortality. In only one way can it do this, and this is by generation, by leaving behind a new being in the place of the old. (208a9–b7)

Built into living things is a passion to reach beyond the self. Creative invention, desire to make good things happen for oneself and others, the impulse to bear children and do great deeds is the spark of immortal transcendence in mortal life.

* * *

How many times Socrates met with Diotima he does not say. As sessions continue, little by little a program is laid out in the form of a progressive widening of erotic energy in vision, thought, and action.

> "Those who are pregnant in body," she said, "turn to women and are in love in this way." In getting children they achieve immortality and remembrance and a state of happiness that they intend to "provide for all future time." (208e3–209a1)

If there is none of Pausanias's contempt here for a vulgar Aphrodite "of the people" presiding over the begetting of children, at the same time having a child is not the end of the work of love. Married couples have great hopes for their children, for the continuation of their name and the happiness of future generations. But simply siring or bearing a child is not enough. Although there is no sharp dividing line between the impulse to mate in animals and humans, humans respond to more than beauty of physical form. Pregnant in spirit as well as in body and inspired by beauty of spirit in others, lovers work to raise and educate their children with energy and purpose that makes both them and their children better persons. In this sense, virtue is an ongoing creation of lovers working together with erotic energy.

> "But those pregnant in spirit, for there are those," she said, "conceive and bring forth what belongs more to the spirit than the body. And what are these things? Mindfulness and other kinds of virtue. And those that produce these are all the creators and makers of things that we call inventive." (209a1–5)[5]

In conversation and in shared activities and pursuits, parents "give birth" to thoughtfulness and intelligence in themselves and in their children. Similarly, creators, innovators, artists of all kinds, erinventors, and discoverers are in love, finding in collaboration useful techniques and new understandings. If the desire for immortality and the good that survives drives on child-rearing, it also drives on literature, poetry, music, inventions, and designs that ease and beautify human life.

Just as there is continuity between the reproductive instinct in animals and the begetting of human children, there is continuity between the nurturing of children and creative activity. No mysterious insertion of immaterial souls into animal bodies is necessary, no previous soul life in some other heavenly world. But nor is the begetting and raising of children mindless biological mechanism. Giving birth, feeding, keeping offspring safe require mindfulness and reasonableness that are products of the spirit. The immortality gained is not the immortality of a soul that hopes to leave the "prison" of its physical body behind like an outgrown coat. It is immortality embodied in natural life in the form of creations and productions that contribute to human happiness. "Birth in beauty" is not limited to the rearing of virtuous happy children or creating beautiful art. "The greatest and the most beautiful mindfulness," Diotima tells Socrates, "is that occupied with the 'management' of households and cities," not "rule" with its assertion of power or adherence to a blueprint model of transcendent form, but literally the making of a "cosmos" of order and harmony in families, communities, and states. "This is the knowledge and skill," said Diotima, "that is called by the names 'reasonableness' and 'judgment'" (209a8–b1).[6]

Success in this "work of love" depends on others as well as oneself. There can be no sudden decision to be virtuous, no act of pure will. The spirit of a child, Diotima tells Socrates, must be "impregnated" *when young* with "reasonableness" and the "desire to set things right." Then, when the child matures and experiences the desire to come together sexually and give birth, she or he will not grasp onto the first available or coercible attractive body, but look for someone to whom to respond with all the awakening and fluttering described by Diotima as erotic arousal in the presence of beauty. For this to happen, a young person must come into contact with beauty. No matter how reasonable or just is an early upbringing, a young person surrounded by ugliness withdraws, sullen and depressed, angry and hostile. Any seeds of reasonableness and justice in her or his nature go dormant, and in such circumstances, even the best of parental guidance may not be enough to ensure the generation of good instead of bad thoughts and actions.

In this sense, virtue is not an attribute of an individual soul, but a joint project generated by lovers joining together in response to more than physical beauty.

> It is beautiful rather than ugly bodies that a person welcomes in his pregnancy, but if he happens also to meet with a spirit beautiful, well nurtured, and good-natured, he happily treasures both in one. Immediately on approaching such a person, he finds an opportunity to talk about what virtues and pursuits are necessary for a man or woman and so takes in hand a kind of education. (209b6–c2)

Now more can come of a relationship with a lover than a pleasurable act of sex or the begetting of children. Lovers become collaborators, encouraging each other in creative endeavor, helping each other become better people.

> For I say the lover, being with the beautiful one, and conversing with him, begets and bears ideas that have been in mind a long time, remembering them both in presence and in absence. And he shares the raising of what has been born with this person, so that they have a much greater community one with the other than comes with children alone and a more lasting friendship, because more beautiful and deathless are the offspring they bear. (209c2–9)

Learning between lovers is mutual. It is not that a teacher or parent with superior knowledge drills a youth in preset lessons, imprinting preformed ideas on a student's mind, or that a lover tries to mold a loved one into a replica of her or his ideal self. Learning goes on as lovers work together, bringing forth long-held inchoate thoughts that are clarified and developed in earnest conversation and remembered later when lovers are apart. Like the raising of physical children, the raising of thought children naturally leads to wider interests in codes of law and systems of government that keep communities together in peace and harmony.

Is caring for children enough to relieve pregnancy of spirit in a man or woman? Diotima's answer is no. Who, she asks, if they had to make a choice between having a child or writing a great masterwork of literature, would not choose to be a poet? Who, if the choice were between having a child and devising a just constitution, would not choose to be a great legislator? She urges Socrates:

> Look at the "children" left by Lycurgus in Lacedaemon, laws which were the salvation of the Lacedaemonians, and I might say of all of Greece. Look at the honor you Greeks have given Solon for begetting the laws, and also to so many others both Greek and foreign, for so many shining good works, and so much excellence begotten. (209d4–e3)

Diotima's progressive lovers understand that the happiness and health of children depends on the beauty of harmonious families and well-governed communities.

* * *

Diotima's examples include achievements of women as well as men. Alcestis is mentioned as an example of valor. "Poets" include Sappho acknowledged as one of the greatest of the Greek lyricists. The teaching and practice of the famous Aspasia, approved by Socrates, echoes in Diotima's references to the management of households and marital relations, so much so that commentators suggest that Pericles's famous advisor and companion might be the original for Socrates's mentor in love. In Diotima's account, both men and women are "pregnant" in body and spirit and both respond to beauty. Both beget and bear children and engage in the generative activities of household management and artistic creation. Not only is there a drive in both sexes to leave behind children, but also a need to "give birth to" other good things. On this view, "giving birth" is not women's work; it is human work, energized by mutual attraction and mutual aspiration. Lovers "give birth" inspired by loved ones, remembering them in presence and absence, "nurturing" and "bringing up" the fruits of their relationships in conversation, children, common pursuits, the creation of good things for themselves and others. Beyond the raising of children, relationships between lovers provide mutual inspiration and discussion that supports further achievements. Epic poems, lyric verse, stirring dramas, ideas, theories, advances in useful techniques, law codes, just systems of government all are fueled by eros, and none are conceivable as productions of a solitary brain.

Although the terms in which Diotima describes the expanding work of love—conversation, interaction, discussion, shared interests, household and community management, civic affairs, artistic creation—are not gender-specific, her examples often refer specifically to the pederastic loves of Socrates's circle. Pubescent boys just leaving the women's quarters approached that masculine society like acolytes entering a Catholic church or young athletes coming under the tutelage of respected trainers and coaches. Not yet with beards or the hard muscular bodies of men, soft and delicate, not ready to assert their manhood, they aroused the desire of men like Hippothales in the *Lysis*, desire that made him "focus all his care, like a slave, on a single boy, or man, or practice" (210c9–d2). These were the troubled relations depicted so graphically in Socrates's "mythic word-picture" in the *Phaedrus*. A man sees beauty in a young adolescent boy. In his mind he makes the boy into a kind of god, whether Zeus emblematic of self-control, or Ares hotheaded and warlike, or Hera regal and queenly.

> Each selects his love from those who are beautiful according to his own disposition. He crafts and fits out the boy like an idol as if the boy was himself a god so as to honor him and celebrate his mystic rites...trying as much as he can in every way to make the boy completely resemble himself and whatever god he honors. (*Phaedrus* 252d5e1, 253c)

At first, as Socrates pictures it, the boy resists. He flees the older lover's advances, "prejudiced" against him, having "heard schoolmates and others say it's shameful to associate with such a lover." In time, flattered at being reverenced like a god, courted by an admired lover, bursting with youthful energy, the boy relents. He accepts an admirer as his special φιλος, or friend and mentor. He keeps company with him. He exercises naked with him in the gymnasium. Desire begins to "swell" and "flow" in the boy, and in such close intimate contact, the "pleasure loving animal" organ of the man rises up. Immediately there is conflict. With a revered agalma or sacred idol, sexual contact is blasphemous, but the boy is flesh and "swelling" with desire. Unknowingly the boy makes it worse. He is drawn to his special "friend," touching and kissing him. Like Alcibiades with Socrates, he finds ways for them to spend the night together. More and more insistently the body of the older lover responds. Soon he is the tortured chariot driver of the *Phaedrus*, struggling to control the conflicting "horses" of his desire. One horse, well trained and obedient, pulls back, understanding that harm to the boy may result. The other, wild and uncontrolled, wants to "leap" on the boy and have his way.

In the man attraction gives way to violent inner struggle. Evil untrained bodily appetite, "crooked in shape, gross, ill-formed, bloodshot eyes" springs forward pulling him toward the boy calling attention to the joys of sex, dragging the good horse of restraint and its struggling driver off course. Drawing near to the boy, the driver has a flash of memory. He glimpses again the god-like idol on its "holy pedestal." He pulls back on the metaphoric reins of desire. The bad horse of the body rears back as the chariot driver fights down physical arousal. "The driver wrenches the bit back, and forces it against the teeth of the wild horse, spattering its evil-speaking tongue and jaws with blood; thrusting its legs and haunches to the ground he delivers it over to pain" (*Phaedrus* 254e2–8). Put in an impossible position, the pederastic lover splits in two becoming his own mortal enemy, his soul at war with a body whose responses he can no longer control.

What is less well represented in the *Phaedrus* is what the boy might be experiencing. In him sexual desire is just blossoming. He has been singled out by a man he admires. Schoolmates warn him. Parents try to keep him safe, sending guardians to see him through the streets. Inevitably with all the flattery, his head is turned. Alone in intimate contact with a revered mentor, his own confused feelings draw him on. Made to understand that there are considerable benefits in having a powerful protector who will sponsor and guide him into male society, he is easy prey. Inevitably he "consents" to sexual contact out of "friendship" or "kindness." Lysias in his speech against love suggested some of what could result. The older lover becomes jealous and possessive, or alternatively slavish and servile. The boy is miserable away from his friends and family. Soon the lover loses interest in him and moves on to someone else, younger and prettier, leaving the boy behind broken in spirit.

Even if his lover is faithful, the boy must play a difficult game, standing back and waiting for his lover at the same time as not appearing effeminate. In pederastic sex scenes depicted on vases and kraters, the boy stands erect and unaroused, studiously avoiding his seducer's eye, embarrassed and in a false position, trying to keep some semblance of dignity. Although vase painters were often reluctant to depict actual intercourse, picturing instead a coupling "between the thighs," the many references to pederastic anal intercourse in comedy and literature make clear that a boy was often in the inferior position of penetrated woman.[7] Nor could Pausanias's ethical proviso—the boy should derive some benefit from his acquiescence—make his position much less difficult.

In Socrates's myth, deliverance comes in a kind of magical transformation. Suddenly, in place of erect male members, nodules of philosophic wings sprout on the boy's and the lover's backs and the two take flight, soaring away from their bodies. Floating high above the clouds, circling with the gods the heavenly spheres, they view objectified icons of eternal justice and knowledge. Even after the inevitable fall back to earth, they retain a memory of these forms, and fortified by dialectical exercise, they are able either to restrict themselves to chaste caresses or, if this proves impossible, engage in sex no more than moderately. In his *Laws*, Plato recommended more draconian legal restraints. Does pederasty promote virtue? Will the "seducer" gain in moderation? Will the "seduced" boy gain in courage? Plato's answer was an emphatic no. The man surrenders to desire and is weakened morally. The boy is forced "to play the part of a woman." Whereas love between equals can be shared and equitable, erotic attachment between unequals can never be reciprocal but is typically "fierce and brutal." Should all forms of pederasty be banned? No, rules Plato, only those that involve sexual intercourse. Limit all sex by law to intercourse for purposes of procreation. Either make sex with anyone but one's wife illegal, or make homosexuality illegal, and if a man goes with a woman not his wife, he will be barred from public office.[8]

Diotima's remedy for the bad effects of pederasty is not Pausanias's ethical remediation, Socrates's flight to heaven in the *Phaedrus*, or the harsh restrictions laid down by Plato in the *Laws*. It is coming to understand the generative work of love. Pursuing a beautiful boy can have no permanent issue in good or happiness. To love only in this way is to be doomed to meaningless repetition of past mistakes and, in the end, death with no promise of immortality.

* * *

Diotima has urged Socrates to look around him and observe natural life: "Have you never noticed the strange and marvelous state that all creatures are in, whether they crawl on earth or fly on air, when they are ready to reproduce?" "Look at the men around you, how eager they are to compete and excel." She has used familiar examples. Every Greek schoolboy learned to admire the deeds of warriors like Achilles and the sacrifices of wives like Alcestis. All had studied Lycurgus's codification of Spartan laws and Solon's reform of the Greek constitution. But now the lessons would enter a new phase.

> In this part of the erotica, perhaps even you, Socrates, might be initiated. But I doubt that you will be able to approach the mysteries and the revelations for which, for anyone who is properly instructed, what has gone before was only preparation. But I will speak of them and I will try my best, and now you on your own part must do your very best to try to understand. (*Sym.* 209e6–210a5)

In previous sessions, Socrates has been "instructed" or "taught." As he approaches the final lessons, he is μυηφείης, "initiated" into a sacred mystery. Instruction must proceed "properly" in sequenced stages one after the other so as to reach the τέλεα and εποπτικά, the "final things" and "things seen," when a proselyte becomes a επόπτης, one having seen. Initiation into the mysteries of Demeter at Eleusis took place over the space of a week, with sacrifices, purifications, and a ritual procession from Athens to Corinth. Initiation into the mysteries of love might take as long as a lifetime.

First, she tells Socrates, it is necessary to fall bodily in love and experience the physical pangs of erotic desire. There is no shortcut here. Beauty perceived in another person, whether in body or mind "quickens" the spirit, makes a lover begin to "swell" with desire to be delivered of new being or a new creation. Just as "*penia*" and "*poros*" come together at a festival of Aphrodite, eros comes into being inspired by beauty. Only when a young person sees beauty can she or he begin to look for it, searching out small buds that might bloom into full flower, sensitive to the quickening of creative response and more and more shrinking away from what is "ugly" in body or mind. In this, surroundings play an important role. Among false images, bad companions, and trivial pursuits, need and cleverness may never come together.

> If a person is to proceed properly in this matter, in his youth he must not only encounter beautiful bodies, but *first* if he is properly guided he must fall in love with one beautiful body and "bring forth" beautiful talk with that person. (*Sym.* 210a5–b1)

The initiate in matters of love does not hide or withdraw from the world; she is out in company, encountering people, mingling and meeting where she experiences and responds to the physical sensations of erotic arousal, the first flushing and yearning for physical contact and creative interaction, the swelling and burning of attraction in response to beauty.

But this is only the beginning of progress in love.

> *Next* the initiate must see that the beauty of one body is "kin to" the beauty of another, and that if she is to pursue beauty of form, it is foolish not to see that one beautiful body is much the same as any other, and love all beautiful bodies, lessening the stress of focusing on just one body, by looking down on that kind of passion and seeing it as a small thing. (210b1–8)

This is not an invitation to promiscuity—not Don Juanism as some of Diotima's critics have charged. Instead, the initiate begins to see a sameness and a limitation in loving one form of purely physical beauty. The Apollo statues representing the ideal young male body popular in Socrates's time were all much the same, as were the slim figures of the kalos boys represented on drinking vases and kraters. The image of the ideal female body made famous by Phidias's nude Aphrodite at Knidos, copied over and over, always had the same seductive curves and graceful features. The generative possibilities of physical beauty are limited. One may admire, touch, grasp, and embrace an ideal body, but little follows. To focus on one shape or form is to become fixated and limited. It is to be "slave to the particular instance."[9]

Now instead of worshiping one unique body or bodily form, dreaming like the besotted Hippocrates of that one body day and night, the lover begins to appreciate beauty of body in a wider sense. She sees commonalities of form and symmetry, harmonies of relationship. Moving on from the violence of erotic obsession with one body, she does not, like the jealous lover described in Lysias's speech in the *Phaedrus*, demand that a loved one be always by her side or try to separate a loved one from friends and family. She sees that the beauty of one body is a small thing compared to qualities of mind and behavior. She comes to value beauty of spirit more than beauty of body, even if the "bloom of it is small." Now she can make the advance to more intimate intercourse. She sees that along with a less than ideal body may go a beautiful spirit, opening up new possibilities for interaction and generative activity.

> Now she will set a higher value on beauty of psyche than on beauty of body. So that even if the flower of beauty in a soul is small, it is enough to love and care for, to bring forth such conversation as will tend to the betterment of the young. (210b11–c1)

The generative powers of erotic desire expand. New kinds of good things result from intercourse with a lover in the form of ideas of virtue and harmonious ways of life. The advance from fixation on bodies to the nurturing of beauty of spirit leads to further expansion.

> It becomes necessary to look on the beauty of practices and laws, and to see that everywhere one thing is kin to another and beauty of body is a small thing. (210c2–7)

Concern for the physical and spiritual well-being of others turns naturally to regard for just practices and just laws, traditions of respect for others, the harmonious management of cities. Ugliness—crime, poverty, trafficking in harmful substances, selling sex—is a threat to the good in any loving relationship. The lover begins to see that the moral and physical beauty of a person is "kin to," related generatively to the beauty of communities in which people live. As a result, there is a natural progression from love of the beauty of a loved one's body and spirit to love of institutions and social arrangements in which bodies and personalities flourish.

A further stage is reached. Along with appreciation of the "kinship" between beauty of body and beauty of mind, and the kinship between beauty in individuals and beauty of social customs and law, comes a widening of love to areas of knowledge.

> From practices she is led on to branches of knowledge and in them the beauty of knowledge. Seeing beauty in so many things, she escapes seeing it in only one person, like a servant, servile and small-minded, who has the care of only one child or person or practice, but rather turning toward the wide sea of beauty and by this spectacle bringing forth beautiful discourses and bountiful philosophy. (*Sym.* 210c7–d6)

Sciences such as medicine or political science bring health and good governance. Mathematics provides both practical methods and models for harmony and order. How can a society be made better? What are the values that should be taught to the youth? How can bodies be kept healthy and plagues avoided? How can wars like Athens's long, futile, deadly conflict with Sparta be prevented? With these questions, love broadens to medicine, political theory, physics, astronomy, philosophy. What is the place of human beings in the cosmos? What are the origins of human life? In none of this widening of response is there rejection of the attractions of personal beauty or loving relationships. None of it requires removal from material existence or leaving the body behind. A whole person, physical and spiritual, continues to respond to beauty in persons, just legislation, and theories in physics. Out on the "open sea" of love, no longer trapped in the small cramped bay or inlet of obsessive fixation, the initiate sails out from care for the well-being of a single loved one to appreciate and respond to beautiful practices and laws, methods of education, and styles of literature, theories of metaphysics. Finally, seeing that all these are "bound together" she finds a voice of her own.

CHAPTER 3

Beauty Itself

Now, with strength and power from what has gone before, comes a certain singular knowledge connected with another beauty. And here, I beg you, she said, you must give me all of your attention. When a person has been instructed up to this point in matters of love, seeing beautiful things in succession and correctly, toward the end appears the wonderful nature of beauty. And this, Socrates, is what all the previous effort has been for.

—Symposium *210d6–e8*

Up to now, the lessons have been practical. Socrates has been urged to grasp the "work of love." He is to move from love of one particular body to seeing beauty in many bodies, and on to beauty of spirit, and to beauty in customs and laws. The movement has been expansive and inclusive, out onto a wide "sea of beauty" in human practices, arrangements, systems of legislation, fields of knowledge. All involved seeing along with acting. But now, Diotima warns Socrates, there is a final knowledge that he may have trouble achieving, a vision at the very horizon of human experience. Beyond the attraction of any individual beautiful thing is the fact of beauty itself that survives any one beautiful thing. This is not a particular physical or mental entity, not any one idea or form; the final "theoretic" vision is comprehensive, ranging over a wide field of phenomena as a glimpse of what sustains and binds them together.

> First it is eternal. It neither comes to be nor passes away, neither grows nor diminishes, is not in part beautiful and in part ugly, not beautiful this way at one time and another way at another time, not in one respect beautiful and in another way ugly, or beautiful from one perspective but ugly from another. Nor will a person see beauty with a face or hands or any other part of a body, or as a particular word or text

> or knowledge, or as existing in some place, or as an animal or earth or sky or any other thing, but beauty existing eternally as itself in one single vision, by itself as itself, ever one vision, all the many beautiful things having a share in it in such a way that although they come to be and die away, it itself is never more or less, but is affected by nothing. (*Sym.* 210e8–211b6)

Humans, animals, natural elements like earth and sky, sacred rocks, inspired teachings, sacred texts, all may be beautiful, all may become objects of love and devotion, but every one of them grows, changes, and dies away. Cosmic forces rearrange galaxies and solar systems. Rocks wear away to dust. Humans and animals evolve and change as individuals and as species. Words gain and lose meaning with use. Nothing mortal stays the same. But in all that creation and destruction, beauty continually reemerges to inspire the generative energy of eros.

The way to this final vision is not by escaping physical existence. There is no contemplative, ascetic, or intellectual highroad to a goddess of truth, no chariot ride with the gods. Instead, the approach begins with responses to physical beauty and the "right kind" of loving practice.

> When someone with the right kind of boy-love moves on and begins to see this beauty itself he nears the final telos. This is the correct approach to matters of love. Beginning from particular beauties, for this final beauty a person must go on, step by step, from one to two, from two to all beautiful bodies, and from beautiful beliefs to beautiful practices, and from beautiful practices to beautiful learning, and on to that final learning of that which is no other than beauty, which is the final knowledge of beauty itself. (211b6–d1)

Loving boys the right way, Socrates and his friends can move past fixation on beautiful young bodies to more generative erotic relations. This is not the intellectual flight of souls as imagined in the *Phaedrus*, nor is it "ethical" in the sense of adherence to a moral principle. Obedience to law can bring one no closer to proficiency in the work of love and the final vision than the dialectical philosophizing recommended by Parmenides. Neither is likely to make permanent changes in a person's emotional responses or pursuits. Equally of little use is the calculated weighing of benefits prescribed by Pausanias in his speech in the *Symposium*. Making sure that pederasty results in some benefit to boys in the form of mentoring and tutoring might mitigate some of the harm done, but it is unlikely to result in generative activities of the work of love described by Diotima. Instead, she urges Socrates to expand his vision and loving, move on from physical coupling to interests in households and civics, and perhaps on to the final vision of beauty itself.

The way to that final vision might have been particularly difficult for the upper-class men with whom Socrates associated. Removed from family life and the raising of children, competing for power and prestige in the overheated arena of city-state politics, sex could all too easily became an act of dominance, of pleasure taken at the expense of someone lower in status. The pederastic lover of the *Phaedrus* trapped in violent struggle between desire and shame at taking advantage of a boy's body and spirit might find it difficult or even impossible to progress in the work of love. As Plato notes in the *Laws*, the generative reciprocity described by Diotima is not possible between a man and a boy. As is argued by Lysias in his speech in the *Phaedrus*, in the pursuit of genital pleasure, frustration is bound to occur, making indulgence unsatisfying and leading to more demands and, in the end, more partners.

The lessons apply to sexual obsession of any kind, but again, Diotima makes specific reference to the pederastic loves of Socrates's circle.

> "My dear Socrates," said the visitor from Mantinea, "living in view of beauty itself is, unlike any other life, the life of a human being. This, when you see it, will overshadow all your riches, fancy clothes, your beautiful boys and youths, whose looks so amaze you and who makes you and so many others, seeing your darling boys and being with them constantly, ready to do without food or drink if that were possible, if only you can see them and be in their company." (211d5–9)

But, although her references are often to pederasty, the unproductive frenzy of sexual pursuit she warns against could as easily be heterosexual as homosexual, as easily experienced by women as by men. All humans are pregnant in body and spirit. All humans long to give birth. All feel the urgency of the first erotic responses to beauty. All are capable of moving on to more generative relationships and finally to a vision of beauty itself. The life of someone proficient in the work of love is not a man's life, or a woman's life, or a noble life, but "the life of a human being" lived to its fullest (211d1–3). Diotima urges Socrates to contemplate the change such a life might make in a person.

> What then do you think it would be like to see beauty itself, pure, clear, unmixed, not filled out with human flesh or color, or any other

> such mortal foolishness, but divine beauty itself in one vision? Would you call that a poor life for persons to live, looking toward it, contemplating it, and having it always with them Don't you think that only then would it happen to persons that, seeing visible beauty, they would bring forth, not images of virtue because grasping onto images, but true virtue. (212a1–8)[1]

Diotima describes the understanding that leads to virtue not as a thought but as something "seen," not as an individual object but as a transcendent aspect of the material visible world that, once glimpsed, changes a person both morally and existentially.

Ethical behavior is often shaped by "images" (ειδωλα) of virtue passed on from generation to generation. For Greek men, many of these images came from Homer, whose epic tales provided models for how to go to war, give a speech, get a ship ready to sail, prepare a sacrifice. Virtue came clothed in the trappings of an ancient Bronze Age culture dimly preserved in folk memory. "*Arete*" was the courage of Ares, god of war; "*aristeia*" was the excellence of the noble class. To be good was to be "*agathos*" with the self-restraint and strategic wisdom of the ideal warrior chieftain. Diotima's true virtue comes not from heroic images but from human experience. Putting aside images and imaginings, a person responds to eternally reoccurring beauty or lack of beauty in persons, laws, and bodies of knowledge. Virtue and goodness are not "forms" or templates by which to measure the goodness of individual persons or things. To trust blindly to such a "form" would only get in the way of the quickening of creative productive energy in response to beauty of all kinds.

Diotima's divine beauty itself is not an objectified concept or icon fixed and eternally existing. It is not one habit of soul, shape of body, form of government, or legislation. These, like everything else in the natural world, change, decay, ossify, become corrupt, beautiful at one time, unjust, ugly, and repellant at another. The courage of Achilles, the righteous wrath of Zeus, doing good to friends and harm to enemies, winning lawsuits, all can be destructive, generating neither good nor happiness. As Socrates often demonstrates in the dialogues, what is called courage can in fact be useless bravado. A friend may turn out to be an enemy in disguise. Repaying a debt can cause harm. Virtuous excellence is not something that Diotima's initiate has, but something she "brings forth" in loving collaborative response to beauty with no guarantee that the result will be perfectly beautiful or good.

Is there nothing, Socrates wondered as a young man, behind the constant generation and destruction of the natural world? Is there nothing, he continued to ask in the *Cratylus*, nothing in all the flux and change of language and experience that one can grasp onto long enough to know with certainty? Diotima's answer to uncertainty and change is a deeper understanding of nature itself. Natural processes of birth, development, growth, flourishing, decay, and death are never ending and inescapable, but inherent in these processes is also emergent beauty as an impetus for the reviving and generating of new forms. Diotima's initiate does not give up physical desire. She does not give up sexual love or the bearing of children. She does not "tame" or repress her bodily impulses, a process unlikely to succeed in lessening their strength. Nor does she wean herself from sense perception by way of increasingly abstract studies in pure mathematics or dialectic as outlined in the *Republic*. If theories in mathematics and science are beautiful, it is not as objects to be admired on heaven's shelf but for creative use in the world, part of a "praxis" and a "way of life" that restores order and beauty that has been lost.

Diotima's references are not to Orphic mythology or Pythagorean number mysticism, but to human life: Falling in love, marrying and having children, expressing oneself and inspiring others in poetry and music, devising methods of management and governing, developing new knowledge in medicine, agriculture, and architecture. In these pursuits, preconceived "forms" or icons are as likely to come from prejudice or tradition as from unmediated perception. It is not that once "the Good" is irrefutably defined, a person can identify particular actions or characters as good. Nor is a well-ordered or just state a state in which a few who know "the Good" rule over compliant masses. Instead, the matrix of the work of love is the familial and civic associations ruled over by Aphrodite Pandamos, Aphrodite of the people rejected by Pausanias. In this sense, the work of love is cooperative generation of harmony that results in good things for individuals and communities consulting and working together.

At the same time, this is not the moral "relativism" so often critiqued by Socrates. What is virtuous or good is not whatever looks right or good to me. Nor is it what is thought to be right or good in any particular culture or society. What is thought to be right can in reality be ugly, an ugliness that the very certain conviction of being right can obscure. As a capability rooted in the need and resource of eros, goodness is the ability to see not only full-blown beautiful achievement but also small beginnings of beauty that might be tended and nourished into bloom. Emergent beauty and ugliness is there to be seen or missed, not out of or above the natural world, but built into natural processes as what drives on striving and purposefulness. The conflicted chariot driver of the *Phaedrus* split in two between genital urges and rational deliberation is unlikely to be able to look beyond the struggle going on inside him. In a more natural state of interaction and continuity between bodily desire and desire for the good, it is possible to make progress in the work of love. Rather than splitting into warring factions, body and psyche are aspects of a single complex person, changing in both body and mind in response to beauty and ugliness. There is no irresolvable conflict between opposed impulses, one bodily and brutish, the other rational and moral, no need for supernatural flight. If desires conflict, it is because interpersonal reality is complex. A desire to have children may conflict with a desire to produce great works of art, which in turn may conflict with a desire to be politically active. If one desire is chosen over another, it is because one rather than the other is seen, rightly or wrongly, as generating more lasting good and happiness and for that reason better achieving the immortality that is the final reward of progress in the work of love. As Diotima puts it: "Giving birth to true excellence and nurturing it into being, this person becomes a friend of the gods, and more than any others, immortal" (212a8–10).

* * *

In the literature of early Greece, there was little talk of any resurrection of dead bodies or survival of souls separated from bodies. "Psyche," or spirit, was an enlivening breath released when the body dies. Prayers to the gods were for health, children, fertile fields, an easy death. By Socrates's time, esoteric teachings had begun to circulate: secret formulae and ritual practices that claimed to be able to release souls from bodies, send them to live with gods, perhaps even become gods themselves. Gods and men do not mingle, said Diotima; now there was talk of ancient writings and "acusmata," "things heard," that could remove that barrier. Where the stories came from is hard to say. Confused accounts of Egyptian burial practices?[2] Strands of Vedic mysticism? Legends of travel to the underworld of the dead? In Plato's dialogues, Socrates repeats many of these stories, but always with a note of caution. Painting his mythic picture of winged souls in the *Phaedrus*, he checked Phaedrus's gush of admiration. Was his speech a "good speech"? Had he done any better than Lysias? Did his speech come from "knowledge" of a subject? Was it based on a clear definition of a subject matter properly divided into subcategories?

If so, he told his young companion, it was only in a few introductory passages where he marked out four kinds of mantic "madness." As for the rest, he said, he had done no more than construct a kind of fiction, which might have helped them grasp some truth, but equally might have sent them off in wrong directions. "For the most part," he told Phaedrus, "the speech was done playfully to amuse."

Later, in the *Phaedo*, in prison waiting for death, Socrates recalled tales of fabulous Orphic cosmology. "Some say," he told the friends and followers gathered around him, deep in the earth is a place of horror with fire, belching mud, and burning torrents of rivers. In that hell, the souls of the dead are judged; those incorrigibly wicked are hurled down forever into the burning abyss; still others are tortured and punished until they repent. Those who lived pure lives are conducted up to a purified earth, and an even smaller elite go higher still up to the ethereal heavens. "Of course," he concluded, "no rational man ought to insist that the facts are exactly as I have described them, but either this or something like it must be a true account of our souls and their future dwellings, given—that is—we can prove the soul is immortal" (*Phaedo* 109–114).

But had they proven it, he and his followers, in the long debate that went before? Had they proven it against all the many arguments against the immortality of the soul made by Simmias and Cebes? Or was the truth of such stories, as Socrates suggests at one point, no more than a "noble risk," valuable in bolstering confidence in the face of death, useful in convincing young men that it is best to be just even though injustice is often more advantageous. The problem is posed bluntly enough in the *Republic*. How in the face of a society ruled by an elite class of men, vying for wealth and power, can young men be made to be good when honest men often suffer? How do you motivate moral behavior when the old assurances of Hesiod—a hard-working man who pays his debts will be rewarded with fertile soil, flourishing herds, and prosperous children—no longer hold true, when cheating a business partner, looting a dependent island for treasure, slandering a political rival brings wealth and power. It could only make it worse that on offer were now supposed ways for the unjust to avoid divine retribution of any kind. Itinerate priests roamed the streets, hawking secret knowledge and advertising "spells and enchantments" to ensure forgiveness for the worst of sins. On sale were purifications to erase depravity, magic formulas to win lawsuits, indulgences and salvation at a price. In this there was no incentive to be good, rather the opposite. Sin and be forgiven. Acquire the magic charm, perform the rite, make the sacrifice, and buy off whatever gods have been offended. Plato's older brother Adeimantus reviews the dubious benefits offered by the "orphic" followers of "Musaeus and his sons."

> Most extraordinary is their way of speaking about virtue and the gods, they say that the gods portion out calamity and misery to many good men, and good and happiness to the wicked. Begging prophets go to rich men's houses and persuade them that they have the power given them by the gods to make atonement for a man's sins or the sins of his ancestors using sacrifices and charms, and with celebrations and banquets. And they promise to harm an enemy, whether just or unjust, with magic arts and incantations that can bind heaven to execute their will. (*Rep.* 364b)

In the competitive press of an imperial city-state, the just man who pays his taxes and tells no lies can seem a fool, and even if one believes in some sort of reward or punishment after death, that too could now be negotiated. A man of means could buy indulgence. He could purchase charms and incantations. If all else failed, he could take sterner measures, become a vegetarian, give up sex, discipline his body, and free his soul. If someone objected—"Even if in this life the wicked prosper, there is a world below in which either we or our prosperity will suffer for our unjust deeds"— he had a ready answer—"Yes, my friend, that will be the worry, but there are mysteries and atoning deities, and these have great power." The message was clear. Sin as you will, knowing that salvation is available. Not only will you not die, but you will be forgiven, that is as long as the proper donations, prayers, incantations, and sacrifices are made (*Rep.* 366a).

There is no suggestion of this kind of immortality in Diotima's teaching. What survives death is not a body or a mind, but beauty and human excellence that human persons create in loving consort with others. If eros is the striving beyond what is given, beyond any single life or single human community, immortality is participation in what surpasses the limitations and dangers of one human life. The mysteries at Eleusis promised an immortality rooted in seasonal regeneration: Persephone restored to her mother Demeter, joyfully reborn in spring with all the seasonal renewal of fertility and new life. The final telos, or mystical vision of Diotima's erotica, reaches past the renewal and fertility of plant and animal life to the generation of human cultural life. All humans, male or female, feel their mortality, feel a pressing urgency to bring forth something good beyond their own limited physical existences. Not only do they need others with which to give birth to what is good, but beauty, "το καλον" to inspire and stimulate that desire.

No one brings forth good things solely from within herself. In Diotima's teaching, there are no innate ideas remembered from a previous life, no rational soul inserted into an animal body by supernatural agency. Only beauty in persons, natural objects, artifacts, works of art, institutions, and laws make it possible for humans to be delivered of the heavy burden of mortal pregnancy. In Diotima's teaching, creative coupling and new birth are the immortal element in the midst of mortal life, built into the natural world as what stimulates joining together. To grasp the transcendence of beauty itself apart from any beautiful thing is to escape death and to achieve an immortality that requires no separation from the body, but comes as a result of long apprenticeship in generative and loving relationships and activities. The initiate sees and responds to beauty in persons, works of art, political systems, and theories of knowledge. She "gives birth" in beauty, relating to others actively and productively. She manages households and performs civic duties, works with fellow citizens for political reform, or contributes to bodies of knowledge. Eventually, she catches a glimpse of beauty itself, not any one beautiful thing, but the immortal pull of a beauty that might come to be.[3] The proof Diotima offers is experiential. "Look," she tells Socrates, look at the mating of animals, look at the desire of people to raise children, look at the passion in men to do great deeds. Look and see how the beauty that inspires these efforts never completely disappears. Join with that divinity in "theophilic" devotion.

* * *

Divinity in this sense had been at the heart of Near Eastern and Mediterranean religious devotion for millennia. Inanna, associated with fertility, sexuality, and enlightened rule, was the most prominent deity of ancient Mesopotamia. Votives associated with Istar or Astarte with similar attributes date in Syria from as early as the fourteenth or thirteenth century BC. In Palestine and Judah, Asherah, worshiped by Hebrew women, was constantly the subject of polemical denunciation by Jewish priests and prophets. In Arabia at Mecca, Allat, part of the triple divinity of Allat, Uzza, and Manat, was housed in the Ka'bah and worshiped at her desert sanctuary outside the city until Mohammad outlawed polytheism in the seventh century. Divinity in this sense is multifaceted, representing spiritual transcendence in many different daemonic manifestations. Deities may be given different names and worshiped in rites and rituals related to particular geographies or cultures at the same time as they provide avenues of daemonic spirituality that transcend boundaries of language and ethnicity. As Heraclitus puts it,

> Deity is day and night, winter/summer, war peace, satiety hunger, changing in the way that fire when mingled what is burnt is named according to its scent. (Kirk and Raven, *Frg.* 203)

In this sense, devotion paid to divinity under one name in no way precludes devotion paid under another. There is no promise of exclusive service and fidelity to one god in one place or, according to one text or tradition, no requirement that temples to other gods be destroyed or that those who worship other gods be exterminated.

In Greece the titles of Aphrodite varied from place to place. As "Harmonia" she fostered living together in peace and friendship, as "Persuasion" the ability to convince others, as "Hegemone" inspired leadership, and as "Praxis" intelligent action. As protector of seamen and voyagers, Aphrodite was quieter of conflict between wind and sea, restorer of peace to conflicting elements. As Pontia, goddess of harbors, she presided facing out to sea. At Athens, she was Pandemos, "of all the people" representing the unification of Attica and the coming together of citizens in deliberative assemblies. Her temple at Paphos on Cyprus attracted pilgrims from all parts of Greece continuously from the thirteenth century BC until the final destruction of non-Christian temples and sanctuaries in the fourth century AD. Her votives can be found all over Greece in tombs and around sacred locations.[4] Virtually every Greek town and village had a shrine or sanctuary dedicated to Aphrodite, either at the entrance to its harbor or above on hills or cliffs.

In what survives of her poetry, Sappho expressed many aspects of this "theophilic" devotion. In one ode, she called on the divine spirit of love to sanctify a gathering of friends on a summer day.

> Come to me from Crete, to this holy
> temple, to your lovely grove
> of apples and altars fragrant
> with frankincense
> Come here where cool water ripples
> under branches of apple trees
> where the ground is shaded
> with roses, where from quivering leaves
> falls down oblivion.
> Come here then, take your seat, Cypris.
> Mingle in our festivities.
> From a golden cup pour out nectar for wine.[5]

The holy temple to which Aphrodite is summoned is a grove of apple trees laden with fruit. Through it runs a brook of clear water with banks shaded by fragrant roses. On a rustic altar burns a lump of fragrant frankincense. The ripple of water, the rustle of leaves, the heady scent of incense mixes with the scent of roses, and distance closes between divine and human, inducing a state of sacred oblivion.[6] Divinity is a spiritual presence invited to a gathering that might take place in one of the goddess's many rural sanctuaries.[7] Grass is green in the meadows, streams run, roses bloom, the humming rustling music of wind and water along with the intoxicating scent of incense bring a moment of transcendence turning wine into the "nectar" of gods. "Friendship with divinity" is consummated in a moment of communion with the beauty of the physical world and passionate fellowship between lovers and friends.

In another poem, Sappho appeals to Aphrodite to help attract a girl resisting her advances.

> Richly-throned immortal Aphrodite,
> Daughter of Zeus, I pray to you
> Weaver of wiles, do not distress,
> Lady, my spirit.
>
> But come here, if ever you did,
> Hearing from afar my voice
> Leaving your father's house
> Golden you came
>
> Chariot yoked, drawn by
> Beautiful swift sparrows on rapid wings
> Came from heaven over the dark earth
> Through the midair. (*Frg.* 3)

The form is a cult hymn with epithets, citation of divine ancestry, and a request for favor, but the diction is light and playful as fits appeal for help from a friend. Again, the sacred setting is outdoors in a meadow or grove. Divinity comes in an epiphany of a flock of sparrows glittering out of golden sunlight in a whir of wings. "Help to bring this girl into my circle of friends." "Make her love me," prayers that echo those made by countless women in small remote chapels.

In a sudden moment of revelation, the poet's voice merges with divine response, crossing the barrier between gods and mortals.

> Quickly you came, and Blessed One,
> a smile on your immortal face,
> you asked *now* what was wrong with me.
> *Now* why do I call?

What is it that I desire so passionately
to have?—"Now whom must I persuade
to join the ranks of your friends, Sappho,
now who is doing you wrong?"

If she flees from you, soon she will pursue.
If she refuses gifts, soon she will give,
If she does not love, soon she will love
Even against her will." (*Frg.* 3)

Patient, but reproving of "mortal nonsense," Sappho's evocation of divine response recalls Diotima's impatience with the romantic fixations of Socrates and his friends. See and transcend the triviality of love narrowly focused on one object. The ebb and tide of persuasion and surrender has a natural rhythm. An unsuccessful lover must move on. Soon she may find herself courted and pursued, the person who resisted her advances attracted "against her will." Beauty is everywhere, unpredictable, never contained in one person or thing, a fact of mortal life to be approached with gratitude and humility. There is no fear in Sappho's approach to divinity. No commanding or avenging Lord demands and threatens. Regardless of divine reproof, the narrator of the poem ends by amusingly repeating her plea for help.

Come to me now, deliver me from
painful anxiety, make all that my heart
desires will happen, happen. You, yourself
be my comrade-in-arms. (*Frg.* 3)

Surviving fragments like "Eros again, brings that limb-loosening bittersweet helpless creeping," universally admired by ancient and modern poets, recall Diotima's descriptions of the physical swelling and fluttering of erotic response to beauty (*Frg.* 464). In one poem, the poet watches a loved one engage in intimate conversation with a rival.

Favored he seems to me like a god,
that man, who sits across from you
listening so closely to your voice
sweetly speaking
and your lovely laughter. This I declare
sets the heart in my breast fluttering.
for when I glance at you, speech for me
is no longer possible. My tongue

stays silent; a subtle
fire steals beneath my skin;
with my eyes I see nothing; my ears
ring and over me comes a cold sweat.
Trembling seizes me everywhere
Paler than grass
I am not far it would seem
From extinction. (*Frg.* 4)

As in Diotima's teaching, the beauty that inspires the physical effects of erotic love is not only the beauty of a well-formed body but the shimmer of spirit from within a body, a "lovely way of walking," "the sparkle of a changing face."

In another poem, a friend about to leave Lesbos cries pitifully on the poet's shoulder, and the poet reminds the weeping girl that what lovers create between them remains "in presence and in absence."

Remember me, whose only care you were
And if no, at least I want to remember
Even if you do not, how sweet
And good was the life we shared.
How many wreaths of violets you wove.
How many garlands of roses you entwined
Sitting by my side.
How many chains of pretty flowers
Made by our own hands we tied
Around your neck. (*Frg.* 7)

As in Diotima's teaching, what is "born in beauty" survives separation. Sappho's lovers carry each other with them, remembering what they have been to each other and done with each other. The "children" brought up and nurtured together by loved ones are moments of shared pleasure, the beautification of everyday life, lyrics to delight, and inspire intimate friends, epithalamia, or wedding songs to celebrate the festive occasion of marriage.

In one of Sappho's wedding processionals, the bride makes her entrance to a lyric fancifully likening her breast to a love nest. "Within her breast, rosy lovebirds make a nest to nurture coming children" (*Frg.* 117). In another, a bride is a coveted apple on the highest branch that only the bridegroom has been able to reach (124). In yet another, the groom is so humorously tall that the roof beams need to be raised by a carpenter so he can get through the door. Along with tributes to the beauty of bride and groom come tender lyrics to reassure a fearful bride as she is conducted to the bridal chamber. As one admirer put it:

> From the poets themselves we can learn how difficult it is to find a melody tender enough to please the Goddess with song. For I think the majority of these poets, even though skilled at love poetry, have pictured Hera with all the courage of boys and girls but left the mysteries of Aphrodite wholly to Sappho for singing to the lyre and the composition of epithalamium. After all the games, it is Sappho who enters the bridal chamber, decorates the room, makes the bed, leads the maidens in, and summons Aphrodite in her chariot of Graces and bevy of loves.[8]

Summoned to the bedchamber of lovers in Sappho's poetry is Diotima's ever-present daemonic spirit energizing material existence and weaving the fabric of creative relationships between friends and lovers, wives and husbands, collaborating artists and musicians, citizens and legislators.

CHAPTER 4

The Spirit at the Center of the World

In Strife everything is in different forms and separate, but in Love they come together and desire one another. From these come everything that was and will be: trees sprout up, as well as men and women, animals and birds, water-spawned fish, and long-lived most honored gods. These are such, but running through each other they become different, so much does mixing change them.

—*Empedocles,* Frg. *355*[1]

Although Parmenides's goddess of Truth warned against the thought of any enlivening "mixing" spirit at the center of the world, Parmenides's contemporary Empedocles, a practical man who participated in civic affairs and practiced medicine, insisted that, regardless of what logicians might claim, change and diversity are experiential facts. Appearances can deceive, but when all one's senses are consulted—sight, hearing, testimony, touch, and taste—it is possible to reach an understanding of physical reality that is "as far as mortal cunning can go" (*Frg.* 342). Nature as perceived, said Empedocles, is endlessly variable, and if that variety cannot have come from one primal homogeneous stuff, it can be understood as generated from combinations of primary elements.[2] As for what sets the process in motion, he identified two primal forces. Strife that mixes up elements and drives them apart, and Love, or φιλότης, that brings elements together in new combinations.

Like Diotima's divine beauty-itself, Empedocles's φιλότης is not a thing—not a kind of matter, or a physical element like water or earth. Nor does it "mingle" with humans so as to be visible with physical eyes.

> Her you perceive with your mind not with dazed eyes, she who is thought to be innate even in mortal limbs. Because of her people think friendly thoughts and accomplish cooperative deeds, calling her Delight by name and Aphrodite. She is seen by no mortal man as she circulates among them. (349)

Philotis is not filled in with flesh or matter, not something that can be perceived with physical eyes, counted, pulled close, held in the hand, or examined. As Empedocles put it, "It is not possible to bring it near in reach of one's eyes, or to grasp it with one's hands—even though it is the primary way in which persuasion enters the human heart" (396). It does not take human form with "a human head on human limbs, or two branches spring from its back, or feet, or flexible knees, or hairy genitals, but is spirit alone, holy, beyond description, darting everywhere throughout the cosmos with swift thoughts" (397). Φιλότης, like Diotima's beauty-itself, is eternally and constantly present, bringing elements together, inspiring new combinations, influencing the thoughts and actions of humans.

In Empedocles's account, the "gentle, blameless and immortal impulse of Love" is responsible for the emergence of life itself. In primal matter under the influence of Strife, the basic structure of the physical universe takes form and elements are separated out by causal mechanisms. Soon, however, in the seething cauldron of Strife, the universal working of attraction begins to generate the variety and mixtures of living forms, countering the destructiveness of Strife with the force of attraction. With Strife, everywhere different forms split apart, but with Φιλότησ, "things come together and yearn for each other" (355). In Empedocles's cosmology, the attraction of love not only brings about the creation of natural kinds out of primal chaos; it fits together the body parts of functioning organisms, creates new species, allows species to flourish and evolve, and is the impulse behind the formation of human societies, traditions, and cultures.

In his speech in praise of love in the *Symposium*, Eryximachus drew on elements of this Empedoclean philosophy. Eros, he said, is a much greater force than human sexual desire and is inherent in all living things including animals and plants. Harmony between lovers is related to balanced organic function in healthy bodies, to ecological equilibrium in nature, harmony in music. Just as in love conflicting elements are reconciled, in music rhythm balances fast and slow beats and pitches are knotted together into melody. The "harmonia" and "persuasion" associated with Aphrodite "dart" throughout the natural world inspiring friendly cooperative relations between elements in the body, between body and spirit, within families, and among citizens.

In Empedocles's poem *On Nature*, Love and Strife alternate and in a sense cooperate. Love draws things together and Strife pulls them apart resulting in a cosmos in a state of creative destructive equilibrium. In fragmentary remains of a later work, *Purifications*, he presented a darker picture. Love's power is diminished. Anger and Strife prevail. Father murders son and son turns against father. In the absence of love comes enmity and war. Men are driven apart beyond all reconciliation. Harsh systems of law are instituted and enforced. Criminals are punished in this life and the next. Anyone who sheds blood, swears falsely, or who "puts his faith in manic Strife" is reborn an animal condemned to an endless round of reincarnation into lower forms of life. He himself, confesses Empedocles, committed violence and went to a place ruled by murder and death (401). He imagines rebirth after rebirth: as a tree, a lion, with only a remote hope of rising through prophet and prince to some kind of immortality.

It is impossible to know what brought about this nightmare vision of a world without love. Perhaps Empedocles underwent some spiritual crisis or fell under the sway of some mantic priest. More likely something was happening in Sicilian towns and villages that persuaded him that the dynamism of φιλότης could be tragically lost. In this later work, when Aphrodite reappears it is only as the distant memory of the benign religious practices of a past Golden Age.

> Among them no War God was worshipped and no Battle Cry, Zeus was not their Lord, nor Poseidon, but Cypris (Aphrodite) was Mistress. Her they worshiped with holy images, with pictures of all kinds of living creatures skillfully done, with the perfume of myriad fragrances, and the burning of myrrh and sweet smelling frankincense, and on the ground poured offerings of golden honey. Their altars were not wet with the shameful blood of bulls. This was thought to be the greatest of defilements, to tear entrails out of living beasts and consume them. (411)

A fall from grace, an overpowering sense of sin, a mortal fear of horrors to come, divine punishment, a world ruled by war and bloody conflict, the horror of eating still warm, sacrificed flesh, and only fleeting memories of a sacred scene reminiscent of Sappho's φιλοτατα celebrating in a summer grove of apple trees. In the world of Empedocles's *Purifications*, Diotima's vision of lovers inspired by divine beauty all but disappears. Human and divine are "out of joint." Strife and enmity prevail as ugliness from which one can only creep away, heavy in spirit.

* * *

Euripides's drama *Hippolytus*, performed in Athens ten years after Socrates received his lessons in love, offered another perspective on the effect of defying the divinity of love. Aphrodite herself sets the stage for the tragedy, speaking from her sanctuary above the town of Troezen.[3]

> The son of Theseus by the Amazon woman,
> raised here by Theseus's father,
> is the only one in this town
> to proclaim I am the worst of Gods.
> He rejects women's beds with contempt.
> He refuses marriage.
> He claims to honor only chastity.
> Every day up in the green forest gorges
> He hunts down wild animals,
> pretending to companions higher than ordinary mortals. (10–20)[4]

Shouts are heard above and Hippolytus rides into the sacred enclosure with a band of his young friends. Urged by an elderly servant to be more respectful—"Why not greet the 'Holy Spirit' (σεμνην δαιμον) there with respect?"—Hippolytus laughs contemptuously. "Me, I don't reverence any gods worshipped in bed." He rides out defiantly, leaving the old man behind to bow down and pray for forgiveness.

> We elders, who must not act like youths
> but must speak thoughtfully and quietly
> for it is necessary to be humble.
> To you, Cypris, we offer our prayers.
> Forgive it if because of his youth a man
> has a prideful spirit and speaks blasphemy.
> Pretend you did not hear.
> Gods must be wiser than mortal men. (115–120)

But his prayer is in vain. Hippolytus's fate has been decided. Phaedra, the young wife of his father, has fallen in love with him and is repressing her passion to the point of fatal illness and thoughts of suicide. When her maidservant tries to save her by acting as go-between, Hippolytus erupts in self-righteous disgust and hatred of all women, and Phaedra, certain that her secret will come out, commits suicide to protect her and her sons' honor.

As events unfold, Euripides accompanies the action with hymns and prayers to Aphrodite. Here her servant warns Phaedra against thinking she can overcome the power of love.

> What profit is there for anyone
> if the only cure for this sort of thing is death.
> Eros when refused comes down on a person in a great rush.
> If you give in, Cypris is gentle,
> but the person who puts herself above others,
> who has arrogance in her heart,
> this is the one She grabs hold of.
> Have you not seen how Cypris
> insults and mocks such a person?
> Cypris is everywhere, in the air, in the sea.
> Love is how everything is born.
> Love plants seeds of affection and makes them grow.
> Out of love very person on earth has come to be. (441–450)

With Phaedra determined not to give in to passion even if it means her death, the chorus of village women pray to never to have to defy the force of erotic attraction.

> Eros. Eros. How from the eyes
> drips down yearning bringing sweet joy
> to the souls of those you attack.
> Not me. Never to me bring such trouble
> but come in measure.
> For no fire or shaft of starlight
> is so powerful
> as that of Aphrodite, thrown from
> the hand of eros, her divine offspring. (526–535)

Shocked at Phaedra's talk of death rather than dishonor, they chant a hymn.

> You, Cypris, unbending of deities
> lead mortals on. You
> on your many-colored wings
> embrace with your swift feathers.

> You that rule over the earth and
> over the crashing salt sea,
> you, who stroke with golden-lit wings,
> you stir the maddened heat of
> mountain wolves, All that live
> on earth or on sea you preserve.
> Over every creature under the burning sun.
> you Cypris, revered by mortals in solemn assembly
> Queen of Heaven, you alone rule over all. (1268–1280)

Informed that Phaedra has died by her own hand leaving behind a note accusing Hippolytus of sexual impropriety, Theseus erupts in a rage, banishing his illegitimate son from Attica. When Hippolytus is brought back mortally wounded, a rival divinity appears in the form of a virgin goddess Artemis who absolves Theseus of guilt and promises Hippolytus immortality. Putting the blame for what has happened on Aphrodite, she declares eternal enmity with anyone "friendly" with the goddess of love. She consoles Hippolytus as he dies:

> Wicked Cypris planned it so,
> lacking honor from you,
> and vexed by your self-control . . . ,
> So be it. But you will not lie under earth in darkness
> dishonored by the will of Cypris in anger
> coming down on your body,
> you so pious and reverent.
> In revenge for this, into the heart
> of mortals most friendly her I will send
> an inescapable shaft. (1400, 1416–1422)

The Theseus of myth had never been a "friend" of Aphrodite. For him women were loot and pleasure, pawns to be used and discarded. He seduced Phaedra's sister Ariadne into betraying her father Minos and, on the way back to Athens, deserted her on Naxos where she died of despair. He convinced Hippolyta, Hippolytus's Amazon mother, to turn traitor to her people and then deserted her for a politically advantageous marriage with the Cretan princess Phaedra. When Hippolyta broke into the wedding ceremony, he called on his guard to kill her. In his most scandalous adventure as a young man, he and a friend abducted the beautiful child Helen and by some accounts fathered on her the daughter Iphigenia later brought up by Helen's sister Clytemnestra (Pausanias, *Description of Greece* 22; Plutarch,

Theseus 33). There is no account of Theseus ever falling in love with the beautiful body or soul of any woman or man. In *Hippolytus*, his expressions of grief when told of Phaedra's death are perfunctory and immediately give way to irritation at his bad luck. Convinced by Artemis of Hippolytus's innocence, he repents, but only out of fear that he will again be blamed for hasty judgment.

Euripides leaves his audience to speculate as to why a young man like Hippolytus might become a sworn enemy of love, claiming allegiance with a "higher" purer form of divinity. In the play, Theseus attributes his son's vow of celibacy to the influence of the faddish "new age" Orphism popular in Athens.

> So now, you boast and drive a bargain
> with your vegetarian diet,
> and for some master hold some
> kind of Bacchante rituals and revere
> some sort of nonsense holy writings. (953–956)

Esoteric secret writings, vegetarianism, reincarnation: what might attract a young man to such cults? Hatred and shame at his illegitimacy? Resentment at being banished from Athens so as to be out of the way of his father's legitimate sons by Phaedra? Distaste at the ugliness of his father's sexual exploits? Euripides also hints at changes occurring in the religious sensibilities of Athens. His chorus of women laments:

> In vain, in vain, by the Alpheus
> and in Apollo's Pythian shrine
> the blood of Greece's oxen pours out.
> while eros, tyrant over men,
> key to the Aphrodisiac
> beauties of intimate bed chambers,
> not honored,
> destroys, bringing all
> kind of calamity
> to mortals, when it comes. (536–545)

Later readers would find other Christian meanings in the drama. Hippolytus would be a sacred martyr on the way to eternal life in heaven. Theseus, tricked by a sinful wife, would be forgiven his anger by a virgin spirit similar to the Christian Mary. Phaedra would become the proverbial weak woman, prone as women are to sexual temptation and satanic influence with the worst condemnation reserved for the maidservant plotting and scheming, promoting the cause of a demon Aphrodite out for revenge.

* * *

> Mother of Aeneas, delight to men and gods,
> nurturing Venus. Under the turning constellations of the sky,
> whether it is the ship-bearing seas you visit,
> or the fruitful earth, by you all kinds of living things
> are conceived to see the light of the sun.
> You, Goddess, at your coming winds flee,
> clouds leave the sky. For you the sweet clever soil
> sends up flowers, for you the surface of the sea laughs,
> calms, glows with soft radiance. (Lucretius, *De Rerum Natura* 1.1–9)

Several hundred years later when Lucretius composed his great Epicurean poem on the powers of nature, Athens was no longer an imperial power. Rome was ascendant, first as a Republic, then, as faction fought faction for power, a dictatorship. Sulla took over as Consul and massacred hundreds of its citizens. Civil war broke out. Ease and comfort disappeared. No one felt the misery more than Lucretius, who in his poem urged the revival of Epicurean philosophy as antidote to the Strife-dominated vortex of factionalism and imperial ambition that was tearing Rome apart. An end to irrational superstition, a rigorously naturalist physics, retreat to private life, enjoyment of modest pleasures—these, insisted Lucretius, were principles enlightened men had to learn to live by. There should be no more "dead weight" "of superstition, mother to sinful and impious deeds" (1.63, 1.83). No more fear of death to distract a man from the rational conduct of life, no more ghosts or bullying gods. No thunderbolt of Zeus striking down just and unjust alike in arbitrary exercises of superior force, no universal Stoic Logos of the cosmos, no magical Orphic life after death to be won by abstaining from sex and eating meat. Search out modest sustainable pleasures, learn to accept the painless nothingness of death.

Given this message, Lucretius's opening invocation to Venus, the Roman Aphrodite, presents something of a puzzle. Lucretius calls on a deity, a deity he continues to cite in worshipful terms in stanzas that follow. He evokes the delight of the coming in spring of Venus in tones reminiscent of Sappho's lyrics. He describes Diotima's animals energized with mating and raising their young, her sparrows swooping down in passionate epiphany. Hearts are "pierced" with love;

animals frisk and play in woods and fields. Everywhere love pulses with passionate longing, inspired by the powerful beneficent goddess of love.

> You, you alone are the governing power of the universe
> Without you nothing comes to be in the shining sunlit world
> to grow in joy and loveliness. (1.22)

Venus is Mother of creation. She is the spirit at the center of the world, sending male to mate with female. At her coming, all nature smiles. She brings the light of Joy, the pleasure of love, and the beauty of peace. She is beauty-itself, the only power that can bring peace to a war-torn world.

> Grant that this savage war on sea and land be quelled
> for you alone have the power to bring to mortals the blessings of quiet peace.
> On your breast, Mars himself, Supreme Commander in the brutal business of war,
> flings himself down, subdued by the incurable wound of Love. (1.34–40)

So Lucretius prayed that *philotis* would overcome Strife, that the fierce god of war, pierced by the "swift darts" of eros, might finally come to rest in the arms of the goddess of love, and Rome could be at peace.

Immediately after the invocation to Venus, Lucretius begins his exposition of the antitheist Epicurean teaching to which most of his long poem adheres. There are no gods. Only atoms, combining and recombining according to mechanical principles. All events are explainable by physical causation. If there are immortal beings, they are elsewhere and far away, unrelated to and unconcerned about humans or human affairs. But in those opening stanzas, Lucretius seemed to give the lie to any purely materialist view of the natural world. As one translator and commentator put it:

> The opening of any long work of literature is bound to be important in our interpretation of the whole . . . When we look at the opening of this poem, however, we immediately encounter problems that concern the whole work . . . How can Lucretius dare to pray to the goddess Venus to become his ally in composing verse, when the whole thrust of his anti-theological arguments later on will deny any possibility of the gods intervening in human life?[5]

Attempts to explain away the apparent inconsistency have been many and various: Venus is no deity but only personified laws of Nature that order the movements of atoms. Lucretius meant the proem simply as a metaphor for the physical world. Lucretius takes poetic license in order to popularize the Epicurean teaching that one should take pleasure in Nature. His description of the goddess's joyful coming in spring is a fanciful way to evoke the Epicurean experience of ataraxia, the peaceful "undisturbedness" that comes when one accepts death as a natural end. Venus is simply physical pleasure, the primary goal of Epicurean ethics. More speculatively, Venus is a symbol for the mythical "Mother of the Rome," figurehead of Rome's imperial expansion.

Certainly Lucretius gave no credence to the anthropomorphic adventures of the gods and goddesses of poetic myth. Repeatedly he condemns such stories as empty superstition and morally retrograde. Much of the violence in Rome's collapsing republic he blames on images like the "thunderbolt" of Zeus and the avenging "pitchfork of Poseidon." The thought that one could win favor for a cause with blood sacrifice and powerful allies in heaven only fuels hatreds that a philosophical approach to life is meant to calm. Hope for divine intervention and fear of divine retribution cloud judgment and agitate emotion. But Venus, as Diotima's inspiring beauty-itself, seems to survive this indictment of religious superstition as an integral part of the natural world, with the power to reconcile enemies and restore harmony.

Later in Lucretius's poem, a different Venus appears. The beautiful peace-bringing Venus of the opening invocation is gone. The "wound" love inflicts is no longer sweet. The impulsive movement of attraction to beauty is now a sign of mortal injury.

> The body is drawn to the source from which the mind was pierced by love
> just as a wounded man falls in the direction of his wound and blood spurts out
> bespattering the enemy who delivered the blow with a crimson stream...
> That drop of Venus honey that first drips into our heart is followed by icy heartbreak. (4.1050–1060)

In long passages of invective, Lucretius vilifies Venus as a carnivorous horror. A man in her clutches is drawn to an object of desire against his will. Haunted by images of the woman he desires, he cannot stand to be away from her. Day by day his madness and his grief increase until they become unbearable. When he finally manages to get access to the loved one's body, the sex act is a violent struggle, more painful than pleasurable for both partners. Never does orgasm satisfy; immediately after sex the lover feels unfulfilled, wants more, longs for some new impossible fusion of bodies. Always he is disappointed. "A pretty face or pleasant complexion gives the body nothing to enjoy but insubstantial images, which all too often pathetic hope scatters to the winds" (4.1093–1096). Inevitably, jealousy and suspicion replace love. In the end comes hatred of all women as evil and deceiving.

What makes matters worse, storms Lucretius, is that the beauty of the woman on which a man's love fastens is never real beauty, but only a deceitful semblance of beauty. In a spate of misogyny rivaling Hippolytus's speech against women in Euripides's tragedy, Lucretius describes the abject female ugliness with which besotted men fall in love. Men fall in love with "black girls, stinking sluts, stunted runts, massive dragons, blazing gossips, and fat girls with enormous breasts, and worse," all due to a woman's ability to mask her ugliness with make-up and fancy dress. Lucretius's final advice to his male readers turns on end the teaching of Diotima. To guard against the pain of eros, a man should use sex only for casual physical release or for the begetting of children, avoiding at all cost "falling in love" with any female body or soul. If a man fails and finds himself "in love," he must seek remedies. He can reduce the tension by masturbation or by engaging in "promiscuous" sexual relationships. He can try to fall out of love by concentrating on, searching for, dwelling on a woman's faults. A woman's natural defects—poor skin, smelly parts—typically covered up with make-up and perfumes can be exposed. A man can penetrate behind the scenes, get into a woman's bedroom and dressing room, and observe the "truth," understanding that if ugly women find lovers it has nothing to do with the "prick of cupid" but is due to her wiles in dressing seductively and pretending to accommodate herself to a man's desires. In the end, the best choice for a man who cannot stay away from women is a marriage of convenience. Find a woman useful and tractable, a woman who is not barren and willing to strive to serve a man's needs, and some sort of marital peace might be achieved.

> Often a woman herself by her actions, humoring
> a man's fancies and keeping herself fresh and smart,
> makes it easy for him to share his life with her.

> Love can be built up bit by bit with constant usage
> For nothing can resist the continuously
> repeated impact of a blow. (4.1081–1085)

What passes for love is routine, a wife there when needed, the "blow" of her irritating presence dulled by constant use.

Some of the divinity of Diotima's Aphrodite was gone even from Lucretius's opening invocation. His divine Venus brings fertility, regeneration, and joy, but there is no mention of the higher stages of initiation into the "mysteries" of love. By the end of the poem, even the joyfulness of fertility has disappeared. Eros is a wound inflicted on men, enticing and tempting them to folly, illusion, insanity, and disease. The goddess is a mistress of demons, a source of frustration and violence, humiliation and bondage, a scourge to mankind. As Rome collapsed into decadence and civil war, Lucretius described its downfall in manic terms. Like dogs, men mount women for orgiastic relief. Women roam in raging heat. Couples cling together in mindless copulation. To fall in love is now the worst of fates. Ares's besotted idolization of Venus allows her to take advantage of his weakness and humiliate him. Venus teases and wounds; she puts her victims in chains. Her smiling arrival in spring is a cruel joke, bringing only vanity, jealously, and anger as sexual desire is reawakened. Not even a memory remains of Empedocles's gentler age.

* * *

Some years later, with peace and order restored under the imperial reign of Augustus, Plutarch in his "Dialogue on Love" returned to Diotima's lessons in love. The setting was an athletic and music festival at Thespiae in honor of Eros attended by Plutarch and his new wife, who are eager to celebrate their union and escape quarreling between their respective parents. At the festival, they are drawn into a dispute. Two men are eagerly competing for the favors of a "kalos" youth, Bacchon at the same time as a wealthy widow is trying to win the young man's hand in marriage. Bacchon's relatives, including his cousin Daphnaeus, are in favor of marriage with the widow. Others argue vehemently for the pederastic claims of the boy's male admirers.

Protogenes, one of the most vehement defenders of pederasty, seconds Pausanias's condemnation of any low-life heterosexual Aphrodite "of the people." Marriage and reproduction, he scoffed, have nothing to do with Eros. A man cannot feel eros for a woman. Married love is a "lax and housebound love that spends its time in the bosoms and beds of women, ever pursuing a soft life, enervated amid pleasure, devoid of manliness, friendship and inspiration." Daphnaeus speaks up in opposition. On the contrary, he insists, only in marriage can friendship develop out of χάρις or "caring" as mutual favor and reciprocal affection impossible between a man and a boy. In pederasty either a boy is forced or coerced into compliance, or else, in "affront to Aphrodite," he allows himself to be "mounted like cattle" and used for another man's pleasure. Boy-love, asserts Daphnaeus, is not the legitimate Eros, son of Aphrodite, but a late-born bastard child of darkness, with its origin in the custom of having boys strip naked to exercise in the gymnasium. Men gain entrance, got aroused, go from embraces to copulation, and then take it on themselves to look down on marriage. They "rail at and vilify that great conjugal love which cooperates to win immortality for the human race by reigniting throughout new generations, prone as we are to extinction." The boy love that is the fashion among philosophers, concluded Daphnaeus, is inevitably deceitful. It pretends "friendship and virtue." It boasts of being "philosophic and disciplined." Under cover of darkness, virtue and discipline disappear.

Daphnaeus's critique of pederasty heats up the controversy, inspiring Bacchon's male admirers to further gynophobic tirades. Heterosexual love of women, rages Pisias, is disgusting and coarse. Everyone knows that decent women can neither give nor return passionate love. The newly married Plutarch can listen no more. Of course, there is love in marriage, he tells the disputants. Without erotic attraction, there is nothing to hold a marriage together. As Pisias continues to berate women, news arrives. The wealthy widow has lured the youth to her house and is keeping him there. All signs are she will be successful in winning her suit. Argument dies down and a young man approaches Plutarch—could he explain to the company what is meant by those who say that Eros is a god?—and Plutarch shares what wisdom he has on the subject with frequent references to Diotima's lessons in love.

Eros, he says, is not a newly introduced "foreign" deity like Isis or Mithras, but a core element of traditional Greek spirituality, "invisible" and "engendered by Aphrodite" (87). Eros in the narrow sense of pederastic "aphrodisiazing" like satisfying hunger or thirst brings only satiety.[6] True marital erotic love lets passion "abate" and leaves a glow in the soul. It "opens pores of complaisance and friendliness" that expand response past obsessive desire for a physical body. This, Plutarch makes clear, is not to eliminate "Aphrodisiazing" in the sense of passionate sexual contact between husband and wife. Physical intercourse is necessary to ground a marriage, create community between spouses, cure quarrels that inevitably arise, and open the way to "harmonious friendship." In married love, sexuality, expressed and abated, is the "beginning of friendship and initiation into sacred mysteries" (145). Without that divinity, the activity to which men give the vulgar name Aphrodisiazing is nothing more than a physical act. Focused on a fleeting moment of genital contact, it is something that "can be bought in the street for a drachma." Eros, on the other hand, is a powerful daemonic force, stronger even than Ares. As such, its effect on the spirit is profound and beneficial. Eros makes the spirit "generous," "liberal," divinely inspired. Nor is there any reason, he says, to think that "bloom of the soul" applies only to men, no reason why a portion of beauty, natural excellence, and "the budding of virtue" should not be found in a woman. Contrary to Pisias's argument, in marriage there is more of an opportunity for friendship and reciprocal favor to develop between lovers, given that they share a life in common and the management of a household.[7]

If the goal sought in love is the cultivation of virtue, Plutarch argues, the joining of two bodies in marriage offers incentive not only for mutual kindness and reciprocity but also for virtue in the form of fidelity and commitment. In place of a burdensome regime as in Stoic ethics, in passionate marriage there is a natural impulse to moderation. Eros teaches self-control, decency, and trust by way of passion, not as an alien discipline. In marriage, body and soul remain in harmony. Sexual intercourse is not an unfortunate lapse into immorality, but strengthens the marital bond, whereas in boy-love, eros inevitably becomes degraded bodily love. Daphnaeus was right, says Plutarch. Either a boy is forced and the result is hate and resentment on the boy's part, or the boy consents and is humbled and ashamed because he is seen as feminine. There can be no χάρις or kindness, because there is none of the consent that only maturity can give. There can be no reciprocity because a boy cannot give on an equal basis. In contrast, sexual joining in marriage is not grudging tolerance of another's needs, not the fulfilling of an agreement for the purpose of marital duty or procreation, not a transitory moment of physical gratification, but the true offspring of Aphrodite. Erotic pleasure is "like a seed out of which mutual respect, kindness, affection, and loyalty can grow between husband and wife." This, concluded Plutarch, was the real motive behind Solon's ruling that a husband should make love to his wife, not just to sire heirs but so that "complaints that accumulate in day to day life are erased" and "love flourishes" between spouses.

Pisias and Protogenes's rejection of women and disdain for heterosexual marriage remain the view of some of the company, but in parting, Plutarch remarks a certain irony. What provoked their diatribes against women turned out to be one woman's power and allure. The wealthy widow inherited money. She controls it in her own right. She is able to offer to the *kalos* youth a beautiful well-appointed house, servants to wait on him, and a comfortable life. She has been clever and "resourceful" in getting him to her house, rescuing him from his male admirers, and summoning his relatives so a betrothal can take place. Every sign indicates that the marriage will be consummated and successful. The mistake of those opposing the marriage as Plutarch sees it was not in defending love between men, but in making sexual desire a god in its own right apart from the divinity of Aphrodite. Without divine Beauty, without the prospect of regeneration and creating good, eros is no more than a craving for physical sensation, something easily purchased in the street or "manufactured" by masturbation. Alternately, without the divine seed of erotic attraction and union, there can be no true binding and joining together in love, but only a vague and powerless affection or "agape."[8]

PART II

Lessons Lost

CHAPTER 5

The Highest One

Observe what I command you this day. I will drive out before you the Canaanites, the Hittites, the Perizzites, the Hivites, and the Jebusites. Take heed to yourself lest you make a covenant with the inhabitants of the land where you go, lest it be a snare among you. You shall tear down their altars, and break their pillars, and cut down their Asherim for you will worship no other god, for the Lord, whose name is Jealous, is a jealous God.

—Exodus *34:11–14*

In the teaching of Diotima, divinity is Eileithyia, Moira, Aphrodite. It is birth, fate, and beauty. It is the beauty that inspires the mating of lack and intelligence. It is the mutual desire that creates new forms of being. It is the attraction that leads to regeneration and rebirth and keeps the physical and social worlds in motion. Inanna in Sumer, Ishtar in Syria, Asherah in Canaan, Dionysus and Adonis in Greece, Cybele in Rome—all were different visions and celebrations of a beauty that cannot be quantified, personified, or objectified. Titles and attributes varied from place to place, rites and rituals took different forms, but no one form of devotion took precedence over all others. Local gods and goddesses continued to receive devotion in sanctuaries and temples, with new shrines and altars built nearby to accommodate the deities of newcomers. Especially this was true in Greece where waves of emigration resulted in a syncretic mingling, mixing, and merging of religious traditions.[1] An imported deity might pair with a local deity as Apollo did with the earth goddess Rhea at Delphi. Family relations might be established between divinities in mythical "pantheons" or "theogonies." A person might feel a special affinity with one religious image or ritual, but it was a preference that in no way precluded reverence and respect for other manifestations of divinity. The result was a rich multiplicity of religious observance and practice.

There was no tearing down of idols, killing of priestesses or priests, or desecration of holy sites. Divinity was not equated with any one or fixed number of divine supernatural objects of belief whose existence or powers can be denied or affirmed. Names, sacred images, rituals, and rites were multiple. No one god claimed exclusive authority, as do human warlords, emperors, and monarchs.

It was no different in the highlands of Canaan where small Hebrew communities struggled in competition with other ethnic groups to fend off imperial powers with territorial ambitions. Asherah's fertility might partner with Baal's rainstorms in loving combination; Istar could be addressed in one place as "queen of heaven" and Astarte addressed with the same title in another. The Hebrew tribal god Yahweh could be paired with Asherah on inscriptions. A shrine or temple might house images of several deities, as when King Solomon's foreign wives brought their native religious images into the Jewish temple.[2] Sacred Hebrew love poetry collected in the biblical *Song of Songs* included many references to Canaanite religion. Rural trysts mirrored Inanna's courtship of the shepherd Dumuzi. The future bride comes down "from the peak of Ama'na from the peak of Senir and Hermon," sites of mountain shrines associated with the Syrian goddess. She is called from "dens of lions, from the mountains of leopards" reminiscent of the Minoan "Mistress of Animals." A royal marriage celebration mirrors Inanna's symbolic marriage to the Sumerian King. A repeated refrain, "I urge you, daughters of Jerusalem that you not stir up nor awake love until it is wanted," recalls prayers to Aphrodite sung by the chorus of women in Euripides's *Hippolytus*. But in the sacred scriptural history of the Hebrew people begun during the exile of their temple elite in Babylon and completed in the last centuries before the beginning of the Christian calendar, a different divinity emerged, a unique supernatural Being who "reveals himself" to particular humans, "leads" armies, "speaks" to chosen prophets, "promises" conquest, and, most important, "demands" exclusive devotion. Unlike the divinity described by Diotima, this God does mingle with men, making covenants, expressing anger, promising benefits, and laying down the law.

If Zeus is the Greek god most closely resembling the Hebrew "Lord of Hosts," Zeus played that role with considerably less authority. In Homer's *Iliad*, Zeus presides as paterfamilias over a dysfunctional heavenly family with little more power than mortal lords like Agamemnon or Priam. He promises nothing to the besieged Trojans even when they have been more than diligent in honoring him, and he can do nothing to avert their final "fate" at the hands of the Achaeans. He interferes in battles less as an all-powerful avenging "Lord of Hosts," or commander of armies, than as a harassed aging patriarch losing his grip on power. Stories of Zeus's adventures might play a role in orchestrating ritual, or convey allegorical meaning, but few educated Greeks took myth as literal fact. For philosophers like Heraclitus, Zeus was the law of the universe that ordains the cosmic alternation of elements back and forth from hot to cold, light to dark, hard to soft. For Anaxagoras it was the cosmic mind or "nous" that governs the universe according to reason. With Aristotle, Zeus was a remote principle of metaphysics, the Unmoved Mover existing by necessity to maintain the movements of heavenly bodies. In Stoicism, Zeus was thought itself, the "logos" or cosmic rationality that governs the workings of nature. To do the will of the Stoic Zeus was not to fight battles but to adapt oneself to life in accord with nature and reason as expressed by Cleanthes of Alexandria, in his "Hymn to Zeus."

> O God most glorious, called by many names,
> Nature's great King through endless years the same;
> Omnipotence by whom with your just decree
> all the universe, spinning around the earth,
> goes where you lead it, willingly guided by you.
> So great is the servant which you hold in your invincible hands,
> your eternal, two-edged, lightning-forked thunderbolt.
> that by its strokes all works of nature came to be established,
> with it you guide the universal Word of Reason
> which moves throughout all creation[3]

Powerful though the Stoic Zeus might be, his rational rule did not exclude other aspects of divinity or rule out devotion to other gods.

In contrast, the god of the Hebrew Bible demanded exclusive honor, reverence, and fidelity in return for benefits he promised a "chosen people." When that exclusive worship was not forthcoming, he threatened famine, bondage, and, on occasion, genocide. He parts the Red Sea and drowns the pursing Egyptians, and escaping Hebrews praise not miraculous forces of nature but Yahweh's warrior skills: "he has triumphed gloriously, horse and rider has he thrown into the sea" (*Exodus* 15:1). In battle, Yahweh is merciless, insisting that captives be killed and upbraiding generals if they spare enemy women and children. To aid Joshua at the battle of Jericho, he sends an angelic "commander" to make sure that invading Israelites "devote all in the city to destruction, men and women, young and old, oxen, sheep, and donkeys with the edge of the sword in battle" (*Joshua* 5.14, 6.21). Israel makes the mistake of conciliation, sending a diplomatic embassy to Egypt for help against the Assyrians. Yahweh rages through his prophets. His paramount role as commander has been usurped. He alone must lead them to the rewards he promised (*Deuteronomy* 30). Why were they made to work as slaves in Egypt, why were they defeated by the Babylonians? Why—after they were finally permitted by Cyrus to return to Jerusalem and in the face of local opposition rebuild the temple—were they almost immediately forced to submit to the sovereignty of Greeks, Seleucids, and Romans? There could be only one answer. They had not sufficiently honored their divine Commander. Priests who prayed to Baal for fertilizing rain had to be found out and killed. The "high leafy places" of Asherah where the Jewish women went to pray had to be destroyed.

Judging from the Hebrew Bible, the lure of traditional forms of Canaanite devotion was great. In Judea, repeated waves of religious repression were required to repress local cults especially those associated with Asherah. After successfully repelling the Northern Kingdom's armies, in gratitude the ninth-century King Asa "took away the foreign altars and the high places, and broke the pillars and cut down the Asherim" (14.3). After repelling an Ethiopian attack, he called for a new wave of religious cleansing in Judea and in the hillside communities he now controlled. In his zeal, he demoted his mother from the post of Queen Mother because she "made an abominable image of Asherah." Still, the writer of *II Chronicles* reports, the "high places" of Asherah remained, and Asa's successors struggled to eliminate them with lapses into tolerance as in the reign of Manas'she who "did what was evil in the sight of the Lord, according to the abominable practices of the nations whom the Lord drove out before the people of Israel" (*II Chron.* 33.2). After a defeat at the hands of the Assyria army, Manas'she repented, but still the high places remained, bringing on ever more violent expression under his successor Josiah who instituted rigorous religious "reforms," purifying the temple and calling for a final eradication of all traces of devotion to Asherah.

> In the twelfth year of his reign he began to purge Judah and Jerusalem of the high places, and Asherim, and the graven and molten images. And they broke down the altars of the Baals in his presence; and he hewed down the incense altars which stood above them, and he broke in pieces the Asherim and the graven and molten images…And he burned the bones of the priests on their altars. (*II Chron.* 34.3–5)

Key in maintaining the rigorous monolatry of the Hebrew Bible were prophets who claimed to be able to speak directly to and for Yahweh, reporting infidelities and relaying back commands and threats. When in Judah and Jerusalem, traditional forms of worship persisted, the prophet Jeremiah cried out to Yahweh.

> Do you not see what they are doing in the cities of Judah and the streets of Jerusalem? The children gather wood, the fathers kindle the fire, and the women knead dough to make cakes for the Queen of Heaven. (*Jeremiah* 7:17–18)

The response relayed by Jeremiah was menacing.

> Therefore this is what the Sovereign Lord says: My anger and my wrath will be poured out on this place—on man and beast, on the trees of the field and on the crops of your land—and it will burn and not be quenched. This is what the Lord Almighty, the God of Israel, says: Go ahead, add your burnt offerings to your other sacrifices and eat the meat yourselves! For when I brought your ancestors out of Egypt and spoke to them, I did not just give them commands about burnt offerings and sacrifices, but I gave them this command: Obey me, and I will be your God and you will be my people. Walk in obedience to all I command you that it may go well with you. (*Jer.* 7:20–23)

After the fall of Jerusalem to the Babylonians, over Jeremiah's fierce opposition, a contingent of Jewish refugees fled to Egypt where they hoped to live in peace. Settled at the Jewish emigrant community at Elephantine, they participated in rites customary there for centuries. Furious at their desertion, Jeremiah passed on Yahweh's judgment.

> I will take the remnant of Judah who have set their faces to come to the land to Egypt to live, and they shall be consumed, in the land of Egypt they shall fall; by the sword and by famine they shall be consumed, from the least to the greatest they shall die by the sword and by famine, and they shall become an execration, a horror, a curse, and a taunt. (*Jer.* 44.12)

The Egyptian community responded defiantly.

> As for the word that you have spoken to us in the name of the Lord, we will not listen to you. But we will do everything that we have vowed, burn incense to the Queen of Heaven and pour out libations to her, as we did, both we and our fathers, our kings and our princes, in the cities of Judah and in the streets of Jerusalem, for then we had plenty of food and prospered and saw no evil. (*Jer.* 44:16–17)

Yahweh's answer reminded the emigrants of what it meant to be the chosen people of a "jealous" god:

> As for the incense you have burned in the cities of Judah and in the streets of Jerusalem, you and your fathers, your kings and your princes, and the people of the land, did not the Lord remember it? Did it not come into his mind? The Lord could no longer abide your evil doings and abominations which you committed, therefore your land has become a desolation and a waste and a curse without inhabitant as it is this day. (*Jer.* 44.21–22)

The Canaanite Anat, with multiple attributes including fertility, beauty, life, and protective power, had been imported into Egypt as early as the thirteenth century BC as evidenced by a pillar dedicated to "Anat, Queen of Heaven, wife of all the gods" at the sanctuary at Bet'She'an. Documents from the site of the Elphantine community include her name along with Yahoo (a version of Yahweh) and Ashima (a variant of Asherah). Her title "Queen of Heaven" echoed "Lady of Heaven" from which the name Inanna was derived, as well as the Greek Aphrodite "Urania." As in Empedocles's Golden Age, offerings to her were not smoke or fat from burning animal flesh, but cakes ritually baked and marked with sacred images of fertility. Incense burned on her altars as it did in rites to Aphrodite described by Empedocles.

But in the new sacred order mandated by prophets like Jeremiah, there was no need for rites in celebration of Asherah or Baal or other emblems of fertility and generation, no need for the development of erotic love from attraction to beautiful bodies to the love of well-ordered human practices. No need to think about beautifying the natural environment, creating mutually beneficial alliances with neighboring peoples or creating stable regional peace. Under an all-powerful divine overlord, such benefits did not come from human initiative but were gifts won through obedience and submission to one God. He is the "highest one" seated on the loftiest throne, "sovereign over all kingdoms on earth," with the power to "give kingdoms to anyone he wishes and to set over those kingdoms the lowliest of people" (*Daniel* 4:17). His divinity is "high," higher than the starry heavens with their regular motion, shimmering fires, and ethereal lights, higher than the high thrones of earthly Pharaohs and Kings. Hymns praise his power. Gifts and offerings of goods or treasure are tribute to his sovereignty. In place of the exultant upright-arm worship pose of votive figures in local cults in Canaan and Greece, his followers kneel down, clasp their palms together in humble entreaty, cast down their eyes to avoid offense.

* * *

As the conquests and victories promised by the Hebrew Lord of Hosts failed to materialize—the Northern Kingdom annexed by Assyria, Judah reduced to a subject tribute state, the Hebrew leadership forced into exile in Babylon—the "Highest One" took on a new persona, mingling with his chosen people in a more intimate way. No longer was he only the all-powerful "Lord of Hosts," he was a deceived and jealous husband. Had he not "married" Israel, made her his chosen people, brought them out of bondage in Egypt, promised them the land of Canaan? Had he not done so out of "love," graciously singling out an unworthy struggling people? Loyalty is required between a Lord and his subjects: they must not have traitorous dealings with enemy powers, they must pay proper tribute. Disloyalty to a divine husband is not only sedition; it is whoredom. A wife must avoid all contact with other men and even the slightest appearance of infidelity. The chosen bride must never enter the houses of rivals, never seem to "play the harlot after other gods" (*Exodus* 34:15–16).

The prophet Hosea speaking in the voice of Yahweh chastised the Northern Israelite Kingdom, fighting a long losing war with Assyria and collapsing into anarchy. It was not aggression or improper governance that was to blame for their defeats, Hosea cried out, but whoredom.

> My people inquire of a thing of wood
> and their staff gives them oracles,
> For a spirit of harlotry has led them astray
> They have left their God to play the harlot.
> They sacrifice on the tops of mountains,
> and make offerings upon the hills
> Under oaks, poplar, and terebinth.
> Therefore your daughters play the harlot
> And your brides commit adultery. (*Hosea* 4:12–13)

Visiting the sanctuaries of Asherah or Astarte was not just idolatry, it was adultery, daughters turned whore, and wives seduced.

Hosea cried out to the "children of Israel." Warn your unfaithful "mother" of the danger. "Plead with her," he told them speaking in the voice of an all but incoherent wronged husband alternating between enraged fury and bewildered complaint. "Plead that she put away her harlotry from her face and her adultery from between her breasts lest I strip her naked and make her naked as the day she was born" (*Hosea* 2:2–3). Did they think that a betrayed husband would spare the children, when the mother goes to other "lovers," when she is so filled with lust that he has had to barricade with thorns the paths she takes to meet them? He gave and promised her so much—grain, wine, oil, silver, and gold—riches she now offers to others. He will strip her naked. No one will rescue her. He will "put an end to her mirth"; he will "lay waste her vines and fig trees" (2:11–12). He will punish her for the "feast days of the Baal gods, when she burned incense to them and decked herself with her rings and jewelry, and went after her lovers, and forgot me" (2:13). Abruptly the tone softens. God pleads. He will "allure her back, bring her into the wilderness, speak tenderly to her" (*Hosea* 2:14–15). He will remarry her in "righteousness and in justice, in steadfast love and in mercy" (2:19). Even though she has been a "harlot" unfaithful to her divine husband, even when in punishment he has destroyed her and her children, taken away her livelihood and "uncovered her lewdness" in public, still he will lure her back with tender words. Quick to anger, a husband may also be capable of mercy.[4]

With the Northern Kingdom annexed by Assyria and Judah forced to pay tribute, the prophet Isaiah emphasized the compassion of a husband God who forgets his anger and takes his cast-off wife back after having deserted her.

> For your Maker is your husband, the Lord of Hosts is his name.
> And the Holy one of Israel is your Redeemer
> and he is called the God of the whole earth.
> For the Lord has called you like a wife forsaken and grieved in spirit,
> Like a wife of one's youth when she is cast off.
> This says your God: for a brief moment I forsook you,
> but with great compassion I will gather you in.
> In overflowing wrath for a moment I hid my face from you
> But with everlasting love I will have compassion on you. (*Isaiah* 54:5–8)

After brutal punishment come renewed vows. Yahweh will not be angry anymore with his Israel. From now on his love will be "steadfast" and his promise of peace will not be revoked. But inevitably jealous anger surges back. How could you lie to me and so forget me? He asks the people of Israel. "Upon a high and lofty mountain you have set your bed, and there you went up to sacrifice" "You have loved their bed, you have looked on nakedness" (57.7, 8). What is she so afraid of? Could she be afraid of him, her protector? But never should she think that her foreign "idols" can protect her: "When you cry out let your collection of idols deliver you! The wind will carry them off; a breath will take them away" (57:11).

In a rare moment of peace, with Assyria's power broken and Babylon's still on the rise, Jeremiah reminded the Judeans of the price to be paid for "obscenities" in high mountain shrines.

> For long ago you broke your yoke and burst your bonds.
> You said you would not serve.
> Yes, upon every high hill and under every green tree
> You bowed down like a harlot.
> Yet I planted you a choice vine, wholly of pure seed
> How then have you turned degenerate and become a wild vine?
> Though you wash yourself with lye and use much soap
> The stain of your guilt is still before me. (*Jer.* 2:20–22)

To express the uncleanness of women who pay respect to Asherah, Jeremiah used the most degraded image he could think of: a female animal in heat available to any male that approaches, ready to have intercourse even as she bleeds from her womb.

> Look at your way in the valley,
> Know what you have done—
> A restless young camel interlacing her steps
> A wild ass used to the wilderness
> In her heat sniffing the wind.
> Who can restrain her lust?
> None who seek her need weary themselves;
> In her month they will find her. (*Jer.* 2:23–24)

Does Israel think she can return to her Lord defiled as she is by playing the harlot, playing the harlot to the point of sitting by the side of the road like a common prostitute waiting for lovers? Even when He brought down on her drought and famine and still she was not ashamed? What are her prayers beside all the evil she has done?" (*Jer.* 3). God will "lift up your skirts over your face, and your shame will be seen" (13:25–27). Again come moments of relenting. The love of the Lord is "everlasting." He will make her happy again with rebuilding, merry-making, and planting vineyards (31).

During exile in Babylonia with the prophet Ezekiel, marital outrage takes the form of pedophilic romance. Israel is a child bride, an abandoned infant picked up by Yahweh as she lies naked in an open field. The child reaches puberty. One day Yahweh passes by and sees she has sprouted breasts and pubic hair. He mates with her, taking her as wife. He cleans her up, dresses her in fancy clothes, feeds her honey and other delicacies. But the child bride is unfaithful. She takes other lovers, and again Yahweh erupts in murderous rage. She has lavished the wealth he gave her on others. She has worshiped rival gods. In graphic detail, Ezekiel describes her lewd behavior, spreading her legs, admiring Egyptian lovers with big phalluses. She is worse than a prostitute because she takes nothing for her favors. Now only the most violent of punishments can quiet God's fury. She will be stripped naked before all her illicit lovers. She will be exposed and humiliated. A mob will be called to stone her and hack her to pieces with swords. A last outburst recalls her degraded ancestry. She has been contaminated by the people of Canaan, dirtied by her Hittite, Amorite, Sodomite heritage, but even her degraded mother and degraded sisters were never as evil as she has become (*Ezekiel* 23:1–3).

Ezekiel describes the supposed lewdness of the "sister" provinces of Samaria and Jerusalem in pornographic terms. Samaria played the harlot with the handsome horseman, commanders, and governors of Assyria. The sexual appetite of Jerusalem was excited by pictures of Babylonians as handsome horseman, "portrayed in vermilion, girded with belts on their loins, and flowing turbans on their heads." They defile her with lust and God turns away from Jerusalem in disgust just as he did from Samaria (*Ezekiel* 23:14–15). Jerusalem allowed Egyptians to fondle her breasts. She played the harlot with "paramours whose members were like those of asses, whose issue of sperm was like that of horses" (23:8). The punishments for sexual infidelity become more violent. God will bring Israel's illicit lovers down on her; they will cut off her nose and her ears, they will kill her sons and daughters.

> They even sent for men to come from far, to whom a messenger was sent, and lo, they came. For them you bathed yourself, painted your eyes, and decked yourself with ornaments. You sat upon a stately couch with a table spread before on which you had placed my incense and my oil. . . Righteous men shall pass judgment on them with the sentence given to adulteresses, and with the sentence given to women that shed blood, because they are adulteresses and blood is upon their hands. (*Ezekiel* 23:41–42)

The result was a doubling back of objectification. If it is harlotry for Jews pledged to Yahweh to worship Asherah, then priestesses devoted to that and similar deities are harlots as are women who go to worship at the shrines over which they preside. If the true followers of Yahweh are clean, the followers of other gods are unclean, guilty of unbridled perversion and sexual evil, a judgment postdated back to an original covenant with Moses.

> And the Lord said to Moses, "Say to the people of Israel I am the Lord your God. You shall not do as they do in Egypt, where you dwelt, and you shall not do as they do in the land of Canaan to which I am bringing you." (*Leviticus* 18:1–3)

Nakedness in front of family members. Incest. Marrying your wife's sister. Bestiality. The unnatural practice of men lying with other men. These were the sexual abominations of those devoted to rival gods, supposedly condoned and even promoted by foreign priestesses and priests, committed even on the altars of their shrines as lure for Israelites wavering from faith in their God, proving that the peoples the Israelites were to drive out of Canaan were defiled.

> Do not defile yourself by any of these things, for by all these the nations I am casting out before you defiled themselves, and the land became defiled, so that I punished its iniquity and the land vomited out its inhabitants. (*Lev.* 18:24–25)

The message communicated through Hebrew prophets was clear. To flourish and multiply, to be the worthy replacement for the inhabitants of the lands promised to them, Jews had to distinguish themselves from polluted others. They had to maintain purity in devotion to their "Highest God" that distinguished them from other peoples who dirty their altars and have sexual intercourse before going to the temple and even in their temples. Describing the evil of rival cults, two words—"kedeshah" consecrated woman or priestess, and "zonah" prostitute—were often interchanged. Priestesses of Astarte or Inanna were "whores," as was any Israelite woman who strayed into their temples. Foreign women prostituted themselves in temples, fornicating with any man who came along.

Once the desire for illicit sex was seen as luring Hebrew women and men to the worship of other gods, once it was sexual desire that was responsible for bringing down the wrath of God on his people, erotic desire itself took on special significance. Sexual morality played a minor role in commandments of Moses, which included only the prohibition of adultery out of respect for the property of one's neighbor: "Do not desire a neighbor's house, wife, servants, animals, or any other of his possessions." Additions to Hebrew law in *Leviticus* and *Deuteronomy* included restrictions on incest common in many cultures, prohibited sex with animals, forbade sex with menstruating women, and condemned homosexual sex as wasteful of fertility. Although restrictions were more stringent for women, who were considered unclean for long periods after childbirth and during menstruation, rules of sexual hygiene applied to both men and women. *Proverbs* added a note of practical caution. Adulterous desire should be avoided because it could lead to a bad reputation and perhaps a man having to pay reparations (6:29–35).

Once conflated with idolatry and sacrilege, sexual prohibition took on new significance. If it was sexuality that lured the Israelites away to foreign idols and brought down on them the wrath of Yahweh, then sex itself was idolatrous, tinged with forbidden reverence. The eroticism of the *Song of Songs* was suddenly suspect. Did it not recall the blasphemous marriage between Canaanite Asherah and the Hebrew Yahweh? Did it not carry a residue of the old mystery rites in praise of generation and fertility? Did it not suggest the ecstatic romance between Inanna and her shepherd lover Dumuzi? In response, priests and rabbis gave the verses new meaning. It was not the ecstatic coming together in love of a beautiful woman and man that was celebrated, but the marriage of Yahweh with Israel. The *Song of Songs* was included in the final version of Hebrew Scriptures, but its secular use at weddings and festivals was forbidden.

* * *

Meanwhile in Alexandria, the Jewish scholar Philo was finding in Greek philosophy new images for the "highest One" of Hebrew Scripture. Zeus's traditional epithets Ο Θέος υψιστος, "the highest God," or Ο Θέος μέγας, "the great God," were used to good effect in the Greek *Septuagint* translation of the Hebrew Bible, suggesting that the Stoic version of Zeus as the divine Logos or Reason might provide a more elevated image for God than jealous husband or angry warlord. Accommodations had to be made. The Stoic Zeus did not communicate with individual humans as did the Yahweh of Scripture with his prophets. Stoic ethics promoted virtue as a means to human happiness rather than obedience to heavenly commandment. The Stoic Logos did not interfere in the natural course of events with "miracles" or "signs" as did the biblical Yahweh, which meant that any biblical narratives that could not be interpreted allegorically would have to stand as evidence of God's power to suspend reason and natural law.

The Pythagorean account of creation that followed Critias's tale of ancient Athenians in Plato's *Timaeus* also proved useful. Taking the floor after Critias, Timaeus told a story of a divine Creator or "Uniter" who brought order to the material world and on creation he gave welcome detail. In the biblical *Genesis*, God said simply, "Let there be light," and light and the cosmos came into existence with no account of how or in what order. God simply commanded and saw that what he willed was beautiful and good. In Timaeus's story, the divine Constructor of the cosmos creates the physical universe in a series of precise mathematical operations. Creation is no arbitrary act of will, but logical and necessary, based on "forms" that serve as blueprints or models in the maker's mind. Timaeus's God was not a conjurer but a sophisticated geometer, arranging spherical shapes and circular orbits of heavenly bodies. Elements were volumes bounded by surfaces subject to the rules of geometry, with each element a geometric figure formed out of the "most beautiful" of figures, from which were generated not only elements but interactions between elements, change and motion, heat and cold, as well as bodily sensations like color and taste. This Creator did not simply command, he worked a compass and protractor, "harmonized by proportion," divided up planets in regular numerical intervals, manipulated various triangles to create space and volume. In this way, he constructed a world of essences and necessary truth, a world "good" like himself, a mirror of his rational mathematical mind because, as Timaeus put it, "nothing that is mindless can be more beautiful than mind"(30a–b).[5] Philo speculated. Might the Hebrew Scriptures and the *Timaeus* have come from the same source? Could Greek philosophy, in fact, have had its origin in Hebrew monotheism transmitted to Greek thinkers by way of the Jewish community in Egypt that supplied the "priests" who, according to Critias's grandfather, instructed Solon?[6]

Again, Philo was forced to make adjustments. In the *Timaeus*, the elements out of which the material universe is made—the "unchanging" stuff of eternal unchanging forms as well as formless matter—preexist creation. This could not be true of the all-powerful God of *Genesis* who created ex nihilo and was not dependent on any preexisting substance. On the other hand, might not the rational forms that were the model for creation in the *Timaeus* be understood as ideas in the mind of God? Was not the "image" in which the *Genesis*

God created man a "form"? And is that so different from the eternal form and design of the biblical tabernacle in the mind of Yahweh, described in *Exodus* as engraved on the mind of every "naturally talented" person as a model?[7] This God was no primitive warrior or lover, he was Reason itself, a philosopher who thinks logically in terms of eternal interrelated forms, a mathematician who understands number, a logician who knows the difference between what "must be" and what "cannot be otherwise." Generations of Christian Platonists followed Philo's lead. Reason in the form of speculative numerology and mystical Pythagorean geometry, retrospectively imprinted onto the physical world in an act of primal creation, would become for Christian Platonists the divinely created form and structure of the universe.

Timaeus's account of creation also cast new light on the vexed question of sexual desire and sexual difference. In *Genesis*, Eve is created in her own right, in one version as partner and helpmate to Adam, and in another version together with Adam with a blessing of fertility and fruitfulness.

> God created man in his own image, in the image of God created he them, male and female created he them, and God blessed them and said to them: Be fruitful and multiply. (*Genesis* 1:27–28)

Timaeus's more detailed account of the created architecture of the human body reflected nicely the danger that women and sexuality posed to divine order. The instructions given by Timaeus's Creator to the demigod helpers to whom he delegated the task of building the first male human body were precise. In the head was to be placed a bit of rational soul stuff donated by the Creator, duly separated by a neck from manly spirit seated in the chest. Lower down, below the midriff in the "women's quarters" were put the animal appetites, less amenable to reason but in principle controllable by the rationally guided will. Only later, as a result of degeneration, came the bifurcation of the sexes and the passions of eros. In subsequent generations, some men, unable to detach their rational souls sufficiently from their animal appetites because of "ignorance or sinfulness," were reincarnated as women with differentiated sex organs. With these new female beings came eros, the "desire and love of the man and the woman," along with the "rebellious male organ that acts like an animal maddened with the sting of lust, and woman's genitals that similarly "swell with desire" (*Timaeus* 69–70). Here was confirmation. If women were attracted to the old cults, it was due to moral weakness in their constitution inherited from a race of degenerate men, weakness that it would be men's task to control.

* * *

Some of the interwoven strands of oral tradition that make up the Hebrew canon recall the creative generative powers of Aphrodite and related Near Eastern deities.

> And God said: Let the waters bring forth swarms of living creatures, let birds fly above the earth across the starry heaven. So god created the great sea monsters and every living creature that moves, and with which the waters swarm, according to their kinds and every winged bird according to its kind. And God blessed them and said: Be fruitful and multiply and fill the waters of the seas, and let birds multiply on the earth. (*Genesis* 1:20–22)

As in the old hymns to Aphrodite, the proliferation and variety of life forms is sacred; fertility and the generation of new life are divine gifts. At the same time, there are important differences. Empedocles's Aphrodite creates as an attractive force inherent in nature. Diotima's Aphrodite creates as inspiration for cooperative productive activity. The Hebrew God creates as maker or craftsman of a world that will be subject to his authority and his rule, and in *Genesis*, that hierarchy of God over man is mirrored in a mandated dominion of man over other living things. Humans are not only enjoined to be fruitful and multiply, but, created in the image of God, they are to rule over other creatures.

> Let us make man in our image, after our likeness, and let them have dominion over the fish of the sea and over the birds of the air, and over the cattle, and over all the earth, and over every creeping thing that crawls upon the earth. (*Gen.* I:26)

Rather than being fertile in its own right, nature is created for man to use as he wills, and in the biblical version of primal history, even that fertility is immediately compromised.

Adam and Eve are cursed for eating the beautiful desirable fruit of Asherah's forbidden "Tree of Knowledge." Eve is condemned to painful childbirth and Adam to painful labor. The beauty of the "daughters of man" attracts the sons of God and human beings multiply and increase, only to be seen as evil by God and wiped off the face of the earth by the great flood. Noah and his wife are allowed to survive, and again Noah is told to be fruitful, but the new fecundity is on different terms. No longer is it the old generative fertility of all living things, but the increase of a purified privileged race descended from one man and his descendants. God renews his order to be fruitful and multiply. He promises that he will never again destroy the human race even if men are evil and consort with "daemons." But along with the postflood injunction to be "fruitful and multiply" comes the mandate for an even harsher rule of man over other creatures. Yahweh tells Noah and his descendants: "The fear of you and the dread of you shall be upon every beast of the earth and upon every bird of the air, upon everything that creeps on the ground, and all the fish of the sea, into your hands are they delivered" (*Gen.* 9:2). In biblical history, divine fertility becomes the fertility of a chosen family line, the result of a pact or covenant made by God with one extended patriarchal family whose special status is marked by circumcision and the observance of purifying ritual and dietary rules.

Much of that purity would depend on the proper conduct of sexual relations as outlined by Philo in the third section of his *Special Laws.* Like Lucretius, Philo began by acknowledging the "power" of eros in language recalling the old hymns to Aphrodite. "No part of the world escapes its domination." "Everyone obeys it with no question." But even though eros is necessary for procreation and is part of God's creation, warns Philo, it cannot be "unnatural" and even if "natural" it must never be "excessive." He reviews the sanctions of *Deuteronomy* and *Leviticus.* No incest. No sex with a woman who is menstruating. No marrying a barren woman. No extra vaginal emissions of sperm. No bestiality. No homosexuality. All a waste of precious fertility. After long passages on what constitutes incest, Philo devotes some of his harshest condemnation to the evil of pederasty. A boy subjected to pederastic sex becomes "effeminate." He has left "no single spark of manly character to be kindled into flame" (*Special Laws* III.37). The seducer wastes his fertility and leads others into sin. The penalties prescribed by Philo for sexual deviation are harsh. For pederasty death (42). For bestiality the same (43). For adultery death (52). For prostitutes stoning (51). Rape of a widow or divorced woman less serious, a lesser penalty for these (64). In Philo's *Special Laws,* worries about the moral and social effects of pederasty expressed by Plato in the *Laws*—the inherent unsatisfiability, the tendency to excess, the humiliation of becoming a sexual object—harden into sexual sin.

In Jerusalem, surrounded by Greek culture and under Roman jurisdiction, Jewish polemics against other cults intensified. If what went on in the high leafy sanctuaries of Canaan had been "whoredom," all the more was worship in Greek and Roman temples, "abomination," "fornication," and worse, as expressed in a passage from the *Wisdom of Solomon*:

> For the devising of idols was the beginning of spiritual fornication, and the invention of them the corruption of life...And this was an occasion to deceive the world: for men, serving either calamity or tyranny, did ascribe unto stones and stocks the incommunicable name [of God]. Moreover this was not enough for them, that they erred in the knowledge of God; but whereas they lived in the great war of ignorance, those so great plagues they called peace. For while they slew their children in sacrifices, or used secret ceremonies, or reveled in strange rites, they kept neither lives nor marriages any longer undefiled: but either one slew another traitorously, or grieved him by adultery. So that there reigned in all men without exception blood, manslaughter, theft, and dissimulation, corruption, unfaithfulness, tumults, perjury, Disquieting of good men, forgetfulness of good turns, defiling of souls, changing of kind, disorder in marriages, adultery, and shameless uncleanness. For the worshipping of idols not to be named is the beginning, the cause, and the end, of all evil. (*Wisdom* 14:12, 21–27)

The translators of the Hebrew Bible into Greek avoided the very term "eros." The word for prostitution, *πορνεἰας*, replaced the Hebrew Bible's "abominations" for practices in non-Jewish temples. "Priestess" and "prostitute" continued to be used interchangeably.

For the Apostle Paul, converted to a new Christian version of Judaism, the divine bridegroom of the *Song of Songs* was no longer the aging Father-God Yahweh but his deified son Jesus. The bride was no longer the Jewish people; it was the Christian Church, pure virgin bride of Jesus as described by John of Patmos in *Revelations*:

> I saw the holy city, new Jerusalem, coming down out of heaven from God, prepared as a bride adorned for her husband...Then came one of the seven angels who had the seven bowls full of the seven last plagues, and spoke to me saying, "Come, I will show you the Bride, the wife of the Lamb." (21:2, 9)

Again came the doubling back of images. A bride must be pure and virgin. She must submit to the authority of her husband. The Church must remain pure, true to Jesus, meaning that proper marital relations between women and men in and out of church must be maintained. Just as the Church is subject to her husband Christ, in the same way a wife must be subject to her husband, cleansed and adorned like the "dark" bride of the *Song of Songs*, so as to be presented to Christ spotless and without "blemish."[8] Just as the Church is part of the body of Christ, in the same way a wife must be absorbed into the body of her husband. This, declared Paul, is the great "mystery," both of human marriage and of the sacred marriage of Jesus to the Church (*Ephesians* 5:21–32). The use of "πορνεία," and later the Latin "*fornicationem*" for devotion paid to other deities in letters and in other gospel writings collected in the Christian "New Testament," would be cited as proof for the all but universal belief among church fathers and later generations of Western scholars and historians that lewd and lascivious sexual rituals and cultic prostitution went on in pagan temples.

CHAPTER 6

Demonizing the Daemonic

> Fallen, fallen is Babylon the great!
> It has become a dwelling place of demons,
> A haunt of every foul spirit
> A haunt of every foul and hateful bird.
> For all nations have drunk the wine of her impure passions.
> And the kings of the earth have committed fornication with her,
> And the merchants of the earth have grown rich with the wealth of her wantonness.
>
> —*Revelations* 18:2–3

In Timaeus's account of primal beginnings, creation takes place in two stages. The divine Creator first orders the material world according to mathematical principles. Then, being immortal himself, he must delegate to helpers or "demigods" the task of building to his specifications the mortal bodies that will clothe the bits of diluted soul stuff he donates for that purpose. Suddenly, in a narrative moment fateful for the future of monotheist polemics, there appeared a race of intermediary supernatural "beings" between God and humans. In subsequent Platonist and Christian writings, "personified daemons" would play an increasingly important role.

In Plato's late dialogue the *Statesman*, a Platonic "Stranger" explains that the activity of demigods continued past the original creation into the early stages of human history when each daemon was given a human "tribe" to look after and guard just as shepherds are charged with particular flocks of sheep. These tribal daemons were "everything" to their people, he said, providing for their needs and accepting their devotions (271d). Similarly, in Plato's *Laws*, daemons ruled over the first human communities as "spirit kings and magistrates," bringing them "peace and mercy, sound law, unbiased justice, and internal concord and happiness" (713d–e). In the *Epinomis*, an "Athenian" expounder of Platonic doctrine elaborates on the different natures of daemons. Below the supreme God, he said, are several kinds of daemonic beings distinguished by the element—fire, aether, or air—that predominates in their beings. Highest are fiery "visible gods" or heavenly bodies; directly beneath them is a race of aethereal daemonic beings, then a lower level of daemonic creatures of the air. Traditional gods like Zeus or Hera have different ranks in the aerial hierarchy, and they, as well other daemonic beings, "demand" prayers if they are to bring good messages (984). Daemons appear in dreams and prophecies. They supervise oracles and prophetic sites. Most important, they are powerful, and the wise legislator should never attempt to eliminate any of their cults. The warning was repeated in Plato's *Laws*. In a well-governed state, existing shrines, rites, festivals, and cults should be left in place to serve the needs of the community and to "further friendliness, familiarity, and acquaintance" between citizens (738d).[1]

The more removed and esoteric became the Platonic One or supreme Form of the Good, the more the channels of daemonic spirituality described by Diotima filled with supernatural aerial creatures inhabiting the space between heaven and earth with ever-expanding characteristics and functions. When Plato's successors in the Academy introduced a more rigorously Pythagorean metaphysic, abandoning all talk of Forms for mystical numerology and cosmic triangles, the roles and functions of "daemonic" creatures further expanded. Cleombrotus in Plutarch's dialogue "*The Obsolescence of Oracles*" cited the Academy's Xenocrates and Chryisippis and others as reporting "testimony from ancient times that in the space between gods and humans there are certain natures that suffer human-like emotions and unwilled changes, creatures that we, like our fathers before us, should see as daemons and pay homage to by that name" (416c).

The problem lay deep in Platonist metaphysics. If there is to be an eternal supreme "Being"—the One, the Good, an Almighty God—existing outside of and independent of the natural world, something has to connect that supreme Being back to the natural world so as to provide a connection between divinity and humanity. Otherwise, there is no reason or utility in any kind of religious practice. All the great variety and diversity of omens, rituals, shrines, sacrifices, and festivals that gave inspiration and joined communities together for millennia would lack meaning and purpose. Few Platonists were willing to take such an extreme position, if not out of fear of the rarely brought charge of impiety, out of a sense of the importance of religious practices in harmonizing, beautifying, and inspiring human existence. Always along with talk of daemons came warnings. Regardless of the superiority of Platonic reason, regardless of the transcendent form of the Good, traditional deities should not be offended or their cults and shrines disturbed. Something had to bridge the distance between the abstract remote One of Platonic philosophy and human existence. Otherwise, spirituality drains from human affairs and religious practice becomes nothing but ignorance and superstition.

Even the most rigorous of Pythagoreans had reason to respect civic harmony, especially on occasions when one of their Brotherhoods attempted to intervene in city affairs.[2] Even dedicated skeptics like Sextus Empiricus, claiming to be able to disprove any argument for the existence of any god, acknowledged the value of "nondogmatic" religious practice and the importance of traditional cults. For these purposes, the philosopher's eternal form of the Good, or One, or equilateral triangle, standing above and apart from the generative motions of the physical world was of little use. How was contact to be made with such divinity by people unable or unwilling to master Pythagorean numerology or mind-bending dialectical exercises, people concerned with the raising of families, the practice of medicine, or the administering of justice in a city? How is any contact to be made with divinity if the eros so intimately engrained in human sensibility is only "*ta aphrodisia*," a degrading itch located deep in the lower reaches of the groin? With divinity removed to a timeless realm of mathematical truth separated from any inspiring Beauty visible and operative in the natural world, daemonic beings were increasingly pressed into service to bridge the gap.

Cleombrotus, in Plutarch's "Obsolescence of Oracles," explains:

> If the air between the earth and the moon were to be removed and withdrawn, concord and community would be completely destroyed, in the middle would be only empty unconnected space. Just as those who deny us a race of daemons make gods and humans remote and alien from each other they would force us to a disorderly confusion of all things. (416e6–f5)

The disorderly confusion, Cleombrotus continued, was to imagine that the deified One of Platonic philosophy could have anything to do with natural life. It was to make the mistake of "bringing God into human emotions and activities, drawing him down into our needs, as the women of Thessaly are said to draw down the moon" (416f5–417a1). At the same time, to say that oracles and mysteries are empty superstition or that religious rites and ceremonies have no relation with divinity was equally intolerable. Rites and rituals like the dressing of the goddess for the festival of Aphrodite in Athens, prayers and donations at sanctuaries above Greece's many harbors and ports, wedding rites that celebrate the divinity of young love, and all the many rituals devoted to traditional divinities could not be simply eliminated. But nor could the sunlit upper world of the Forms in the *Republic* or the divine Geometer of the *Timaeus* have anything to do with such practices.

For this, supernatural servants and agents were called to duty, not only to assist at the first creation of living things and to supervise the tribes of men in the early days of the world but also to "serve" as administrators and guardians of sacred rites, "prompters" of revelation at the Mysteries, and administrators of oracles. What Plutarch calls the πρόνοια or the "Providence" of the Platonic Creator had to make a difference, but in a way that did not compromise the oneness and purity of Platonic Being. Cleombrotus explains the dilemma:

> Given it is hard to grasp and define how and to what extent Providence matters. But those who make the god responsible for nothing and those who make him responsible for everything go beyond what is moderate and fitting. ("Ob. Or." 414f1–5)

Nor could the problem be solved by the simple addition of inert formless matter as the material of creation as is suggested in passages in the *Timaeus*. Much more convenient, said Cleombrotus, is to "set a race of daemons midway between gods and men" ("Ob. Or." 415a2–3).

In support of personified daemonic beings, Plutarch quotes a subtly reworded passage from Diotima's description of daemonic spirituality in the *Symposium* (202e). Instead of Diotima's daemonic "conveying" (διαπορθμενον) between gods and humans, daemons are "serving" (διακονικόν) (*Isis and Osiris* 361c2–5; "Ob. Or." 416f3). The change is significant. No longer is the daemonic power of both eros and religious devotion identified with the "conveyance" of beatific vision in many forms of spiritual insight and practice. Instead, daemonic beings "serve" and potentially "disserve" a powerful master.

Plutarch was quick to see the danger. As daemonic beings proliferated so could "*deisdaemonia*," superstitious "dread of daemons" that can lead to the evil of atheism and the undermining of all religious practice. Obsessive fear of the gods, belief in magic and miracles, taking fabulous stories of myth literally, engaging in secret rites to win divine favor, all, said Plutarch, are evidence of impiety rather than piety. To fear gods is to believe that gods could do injury or evil, a belief that inevitably leads to apathy, immorality, failure to understand, and mockery of religion. To think that special dispensation can be gained in return for incantations, worship, or gifts inevitably ends as an argument for atheism as claims become more unbelievable and recitations, sacrifices, and observances more extreme. And, Plutarch argued, participation in offerings in temples, pilgrimages to shrines, sacrificial meals, participation in festival dancing and singing bring a spiritual sense of sacred calm, delight, insight, and community that no successful civilization can afford to do without.[3]

* * *

In Timaeus's creation story, there is no question whether the Supreme Geometer's demigod helpers would faithfully obey his orders. They are his creatures and work from his instructions. Given independent existence in their own airy habitat and reimagined as suffering from human-like sensations and passions, daemons were soon free to willfully deviate from reason and virtue. Fully personified, they began to take on the vices of men, a view that Plutarch attributed to Xenocrates, as well as to Plato, Chryisippis, and Democritus, who "clearly recognized that there exists another stubborn and wretched race of daemons that are willful and aggressive" ("Ob. Or." 419a10–12). According to the discussion in "Obsolescence of Oracles," some of this invention of evil daemons came from foreign influences, most notably from Persia where Zoroastrians projected a cosmos divided between a material realm populated by evil spirits and a good but threatened kingdom of light. Cleombrotus speculated: "did this doctrine [of a race of evil daemons] come from the wise men of the cult of Zoroaster, or is it Thracian, going back to Orpheus, or Egyptian or Phrygian" ("Ob. Or." 415a5–7). He tells a story of meeting and consulting with a seer near the Persian Gulf who claimed to consort with nymphs and daemons. He repeats travelers' tales of passing by a distant island and hearing the voice of the great daemon Pan. He cites evidence for the existence of evil daemons in violent foreign rites and rituals: the eating of raw flesh, the tearing apart animals, fasting, beating the breast, and obscene language, Such practices, said Cleombrotus, could not have been ordered by any god. What god would demand that kings or generals "hand over their children, submitting to preparatory rites, and the cutting of their throats?" His conclusion was not that such evils, if they existed at all, were the result of human superstition and folly. Instead, said Cleombrotus, practices like human sacrifice "must be meant to placate and appease the anger and oppressive temper of harsh and avenging deities, or some other mad and tyrannical '*erotas*' who either do not have the power or do not have the desire to communicate bodily or through the body" ("Ob. Or." 417d).

Echoes sound of the legend of the sacrifice of Iphigenia at Aulus popularized by Homer and Greek dramatists. There is a hint of polemical denunciation of supposed barbaric religious practices in locales like Carthage or Phoenicia that resisted Roman rule. There are vestiges of the Greek "αλάστορες," avenging spirits that traditionally haunted and hounded those guilty of the murder of relatives. But the new daemons were not the old Eumenides or Furies. The daemonic sense of sacred guilt that haunted tragic figures like Orestes was a natural consequence of horrific acts and consciousness of deep wrong. As described in Aeschylus's *Eumenides*, the Furies "worked through the body," "binding the heart, tuneless, withering moral lives" (345–346). In contrast, the new Platonic daemons gave vent to their own uncontrolled passions, irrational rages, and tyrannical demands. Guilt that once "dogged" tragic heroes like Orestes was displaced onto renegade "agents," to be pacified in bloody sacrifices, or outwitted with clever arguments as in the exoneration of Orestes of the crime of matricide at the end of the *Oresteia*.[4]

How is it that a demigod created by a providential God can be cruel and tyrannical? How can an agent of God do evil? Here Pythagorean stories of the transmigration of souls could be called in as explanation. If one natural element can transmute to another, and creatures of earth can become airy and, conversely, airy bodies can take on earthy weight, then daemons also can change in status. Transmigration can be upward. Men can become semidivine "heroes." Heroes can turn into daemons and even, due to "supreme excellence" and "after purification," rise to the status of gods. At the same time, transmigration can be downward as explained here by Cleombrotus:

> From these daemons a few of the highest excellence, after a long time, and after purification, come to completely share in divinity. With others it happens that they do not control themselves and some are sent down and clothed in mortal bodies and live a shadowy darkened existence like mist or vapor. ("Ob. Or." 415b14–c5)

Such demotions could account for the existence of evil daemons. Daemons who once had bodies necessarily retained some of the corruption of those bodies. As a result, they could become so compromised with matter they are weighted down by sin and vice.

At times in Plutarch's dialogues, the old naturalistic daemonic spirituality described by Diotima resurfaces. In "Obsolescence of

Oracles," the issue under discussion is how to account for the fact that oracles go silent without supposing that a capricious god provides and then arbitrarily withdraws spiritual guidance. But to say that supervision of oracles is delegated to daemonic agents only compounds the problem. Does God order his agents to withdraw from a particular oracular site? Does he decide not to let them bring enlightenment or help to humans after all? Worse, is he incapable of controlling his own agents? After long digressions into ancient hearsay and abstruse mathematics, Plutarch in the person of Lamprias provides a naturalistic answer. If bodiless daemons can foretell future events, why not the embodied soul of the Delphic priestess? Body sensation may cloud spiritual insight, but there is no reason to think it blocks psychic inspiration altogether. If humans have the ability to remember what is past and so no longer present, why might they not be able to envision the future, especially when the influence of immediate bodily perception is minimized as it is in dreams or close to death? "The best of seers," says Lamprias, "are intelligent persons who have in their souls a special sense that along with reasonable probability guides them on their way" ("Ob. Or." 432c11–13). Such a person is receptive to "impressions" and "presentiments," especially if she or he is in a temporary state of withdrawal from the present that clears the way for deeper insight. And this state of prophetic inspiration can be furthered and fostered by natural fumes or infusions that come from the earth as at Delphi, allowing priestesses to receive ever more directly a "prophetic current and breath equally divine or holy whether it comes from out of the earth or from a spring of running water." What is important, Lamprias concludes, is a state of consciousness that activates bodily and mental channels and networks and brings to awareness "stored away and unperceived thoughts" (432e7–8). On this naturalistic basis, the disappearance of oracles is easily explained. Earth and climate change like all of nature. Effusions dissipate, springs dry up. The body or mind of a priestess may be troubled or distracted, keeping her from concentrating on submerged "impressions" and "presentiments."

Ammonius raises the obvious objection:

> A little while ago we gave in to logic and took the prophetic art from the gods and gave it to daemons, Now it seems to me that we have expelled those very daemons in turn, driving them away from the oracle and the tripod if we reduce the origin of prophecy or rather its very being and power to winds, vapors, and exhalations. ("Ob. Or." 435a5–9)

Plutarch, caught between his conviction of the value of traditional religious practice and his professed Platonism as priest at Delphi, reverts to daemons. He would be horrified, he says in the person of Lamprias, to think that he had contributed to atheism or undermined belief in the gods. No, he assures the company, he has not done away with daemons. Of course, the Creator sends aerial supernatural creatures to supervise and regulate and properly maintain infusions and vapors. Whatever the material cause of inspiration, it is the Creator's design and daemonic maintenance of that design that is crucial.

* * *

Stories of a race of daemons living in the airy space between God and humans were passed on and endlessly articulated by generations of Jewish and Christian Platonists. Philo in his cosmology included the demiurge of the *Timaeus* and his lesser daemonic aides and servants. He found retrospective evidence for daemons in the *Septuagint* whose translators used the Greek αγγελος or "messenger" for the Hebrew Bible's mal'ak or divine agent sent by God. The "messenger" sent to tempt Job, the three visitors to Abraham, the appearance of an angel to Moses, the angelic "Commander" sent to instruct Joshua at the battle of Jericho, all were daemons, said Philo, and not all of them were obedient to their divine master. A punitive, accusing, testing messenger might have done Yahweh's bidding in *Job* but an evil "ha-Satan" (obstructer or adversary) acted against God when he advised King David to order a census of his troops rather than rely on faith in the power of God (*I Chron.* 21:1).[5] God, said Philo, created "ranks" of angelic "messengers," invisible helpers of air, who feel emotion, are kind to good men, hate bad men, and are "guardians of nations," with a special protective guardian archangel assigned to Israel as God's chosen people. Other evil daemons are quite capable of disobeying their maker and rebelling against God.

Philo and other Jewish writers made constant use of evil daemons to condemn the worship of rival deities. In the *Septuagint*, the Hebrew word for "idol" was translated as the Greek "daemon." "The gods of the nations are idols" became "the gods of the nations are daemons" (*Psalms* 56.5). Traditional deities were not simply meaningless bits of carved wood or stone. They had a real presence as malevolent supernatural beings, provoking ever harsher warnings against "abominations" in pagan temples. The idols or daemons of the Gentiles were "a snare for the unwise," the beginning of "spiritual fornication" (*Wisdom* 14). In pagan rites, children were sacrificed to daemons in secret and savage orgies. Anyone who attended temples where daemons were worshiped forgot all charity and condoned disorder in marriage, adultery, and all sorts of uncleanness.

Jewish apocalyptic literature popularized elaborate tales of hordes of fallen angels ruled by a Supreme Prince of Darkness. In *Enoch*, Job's mal'ak is chief of a band of rebel angels who tempt God himself into testing Job and ordering Abraham to kill Isaac. Daemons were said to inhabit non-Jewish temples and possess the souls of evil Roman emperors. Cast out of heaven, they went on to cause havoc on earth and would do so until a final cosmic battle is engaged between the forces of good and evil.[6] Christian writers were quick to take up the theme. In *Matthew*, Jesus is tempted by διάβολος, a single powerful demonic person in cosmic opposition to the good angels or "αγγελοι" who come to him after he has repudiated the "evil one." Jesus gained his reputation as healer by casting out "daemonia," evil spirits that take over human souls and bodies. John of Patmos in *Revelations* imagined the coming of the Kingdom of God as a duel between the good archangel Michael and a horde of rebellious daemons led by a master daemon "Satan" doubling as Tiamat the primal female dragon of Babylonian myth.

> Then war broke out in heaven. Michael and his angels fought against the dragon, and the dragon and his angels fought back. But he was not strong enough, and they lost their place in heaven. The great dragon was hurled down—that ancient serpent called the devil, or Satan, who leads the whole world astray, hurled to the earth, and his angels with him. (12:7–9)

In apocalyptic visions, the end of the world would come with the final defeat of Satan and his hordes. The "devil" would be forever cast down, imprisoned in the center of the earth, providing the image later popularized by Dante in the *Commedia* of Satan at the center of the earth gnawing on his victims.

In Christian imagination, devotion to any god other than the Jewish or Christian god was now demon worship. Visiting Athens, Paul is horrified to see that the city is "full of daemonic idols" and proceeds to express his dismay in the Synagogue and in the market place (*Acts* 17:16–19). He admonished the Ephesians: "In the past you walked according to the ways of the world, according to the Prince of the power of the air, the spirit that now works in the children of disobedience" (*Ephesians* 2:2–3). More and more there emerged a Master of evil, the διάβολος, as a full-blown personified rival of God.

Paul lectured the Ephesians: "Put on the armor of God that you may be able to stand against the wiles of the Devil" (*Ephesians* 6:11). Those who questioned the "new covenant" announced by Paul were Satan "in disguise," serving the cause of the Arch Demon (*II Corinthians* 12:13–15). In *I John*, sinners are "of the devil" (3:8) but those "born of God" do not sin because the "bad one" cannot touch them (5:18). Jesus, angry that Pharisees questioned his claim to be the son of God, evoked the Devil, telling them: "Your father is the Devil and you will do the desires of your father" (*John* 8:44).

A King of demons was held responsible for the very origins of evil—"He that commits sin is of the devil for the devil sins in the beginning" (*I John* 3:8), as well as for the worst of sins—"The devil put it into the heart of Judas to betray Jesus" (*John* 13:2). The Devil was an adversary more powerful than any act of human reason or will power. Only with the grace of God and the rites of the Church could anyone hope to resist his evil force. Members of the Christian community in Corinth wrote anxiously to Paul. Should they "eat with daemons" meat that had been sanctified to the pagan gods of the city? Should they share community meals when the meat eaten might come from animals blessed by pagan priests? (*I Corinthians* 8)[7] Paul's response was measured. If pagan idols are nothing but carved stone, eating the meat could not be dangerous. On the other hand, if eating it upset others in the Christian community, unconvinced that pagan gods did not exist, it was better to abstain to spare their feelings. With Christians an endangered minority, with the Christian community split apart by doctrinal and personal differences, with his own leadership in question, Paul had reason to be diplomatic. Later, with Christians facing persecution and death for refusing to respect state gods, Christian rhetoric against pagan demons was more heated. In Rome, Justin warned of fallen angels out to enslave the human race in the person of false pagan gods working through "authorities" in power in Rome who demanded participation in evil licentious rites and killed anyone who refused to participate in their "demonic" revels.

In Christian writings, a special association was made between demons and sexuality. Demons could tempt a man to steal or lie, but even more dangerous was their evil power to lure a man into "*pornea*," literally prostitution but in Christian parlance now a derogatory term for sexual desire of any kind. The more salvation was deflected from the restoration of the historic Kingdom of Israel to life after death, the more the desires of the body were associated with demonic temptation. To resist that temptation was a task so difficult that even the

Apostle Paul cried out in despair that appetites in the "women's quarters" of his groin could not be subdued by his mind.

> I delight in the law of God in my inmost heart, but I see another law in my members that is at war with the law of my mind. leading me captive to the law of sin in my members. Wretched man that I am, who will save me from this mortal body? (*Romans* 7:22–24)

In his speech in favor of love in the *Phaedrus*, Socrates acknowledged some leeway in accommodating the bad horse of carnal desire. Complete escape of the winged soul from its body, desirous as that might be, was not always possible. Pederastic lovers, who did not succeed in remaining chaste but who went on "moderately" with their "Aphrodisiazing," were not irrevocably condemned to an animal existence. Having seen the Good, they retained some virtue and might eventually advance in future reincarnations. Now, with the "law of the mind" the Christian "law of God," any defeat in the battle with desire, if only in thought, became insult to God jeopardizing any hope of salvation. Physical pleasure of the erotic kind was not only human weakness; it was fornication and the worship of demons. Human desire was naturally excessive, a constant demonic presence within the body tempting a man away from any hope of final bliss. In the Christian imagination, satanic beings and pagan demons were constantly on hand to arouse and lure a man or woman into sexual depravity. It was better for a man, Paul told the Corinthians, to be like him and never touch a woman. If that proved impossible, as it often would, a man should marry so that an exchange of sexual service could guard him and his wife against the temptations of "Satan" (*I Corinthians* 7:1–8). No longer a pathway to divinity, eros was now toxic medicine to be taken in small doses to avoid a greater disease.

Paul advised the Thessalonians to take that medicine with a minimum of contamination of erotic desire:

> This is the will of God, even your sanctification, that you should abstain from fornication, that every one of you should know how to possess your vessel in sanctity and honor, not in the passion of desire as do the Gentiles who know not God. (*I Thessalonians* 4:3–5)

He called on the Ephesians to renounce their lustful pagan pasts. "We all in times past had our conversations in the desires of our flesh and of the mind, and were by nature the children of wrath even as are others" (*Ephesians* 2:3). As hope faded that the "end times" and the ascension of the elect were near, celibacy became proof of readiness for life after death. What could be better than to make oneself a eunuch for God? What would better prove devotion to God than to deny the power of the old daemonic eros? The very word "eros" was anathema. Avoided in the *Septuagint*, dropped from vocabulary in the Gospels, it disappeared from Christian use. The celebration of loving and fruitful sexuality prominent in the myths, scriptures, iconography, and rituals of Near Eastern and Mediterranean religion: Sumerian love poetry, worship at the high leafy sanctuaries of Asherah, the coming of Aphrodite Pandemos in Athens, the ritual bathing and dressing of Venus for the spring festival in Rome, all were demonic perversion, shameful lust, emblems of ungodly rituals that went on in pagan temples.

There exists no trustworthy evidence of any obscene practice in non-Christian temples. No credible non-Christian source reported sexual acts or "cult prostitution" as part of the worship of any Near Eastern deity or any Greek and Roman deity. Herodotus, notoriously unreliable when it came to hearsay accounts of foreign practices, passed down one not very credible story of supposed ritual deflowering outside a Mesopotamian temple.[8] The travel writer Strabo on a visit to Corinth passed on a tale he heard there about a temple to Aphrodite in ancient Corinth so extravagantly wealthy that it "had more than 1000 slaves and courtesans" (*Geography VIII* 6.20). No part of ancient Corinth or its temple—which archeological ruins indicate was to Apollo and not Aphrodite—survived the Roman conquest 500 years before. All of its original inhabitants had been killed or expelled. Nor was Strabo's story, as it stood, evidence for temple prostitution. Ancient Corinth was a busy seaport and trading center filled with sailors and travelers. As a matter of course, prostitutes practiced their trade in the vicinity of temples to which they might have donated part of their earnings.[9] It is not clear even from Strabo's remarks that donated slaves played any role in ritual. More credible in light of archeological remains and historical evidence was Strabo's description of the existing mountain sanctuary to Aphrodite above Corinth on the summit of Mt Acrocorinthus near a spring "of remarkably fresh cool water."

Nevertheless, Paul and later Christian writers had no doubt that unspeakable sexual acts went on in pagan temples. Warning the Romans about the pagan practice of seeing divinity in persons and things, Paul immediately added a charge of sexual misconduct.

> [Unbelievers] change the glory of the incorruptible God into an image made like corruptible man, and into birds, and four-footed beasts

> and creeping things, Wherefore God also gave them up to uncleanness through the lusts of their bodies to dishonor their own bodies between themselves...and God gave them up to vile affections for even the women did change the natural use of the body into that which is against nature, and likewise the men, leaving the natural use of the woman, burned in their lust toward one another. (*Romans* 1:23–24)

As in the Hebrew Bible, the worship of other gods was not only idolatry but, as a matter of course, sexual misconduct, unlawful homosexual intercourse, lack of sexual restraint, and vile practices,

As churches grew in number, and ecclesiastical offices multiplied, ever stricter continence was required of deacons, priests, and bishops. Just as Jesus's apostles left wives and families to follow Jesus, it was best, said Paul, if married priests and bishops no longer had sexual relations with their wives. Better still would be to come to the priesthood unmarried, never having tasted the evils of sex. Second- and third-century Gnostics and Christian Platonists followed suit. In his "On First Principles," Origen endorsed the cosmic vision of the *Timaeus* along with Pythagorean theories of the soul's preexistence, the struggle of the human soul to be free from the body, reincarnation, and, of course, daemons. Daemons, he said, crave and feed on the blood and scented smoke of pagan sacrifices. They do tricks of magic and prophecy. They inspire civil authorities to persecute Christians. They band together in demon armies, whose deployments he went on to map in elaborate "demonologies." Demons tempt men in the guise of beautiful women; they put evil thoughts in men's minds. Victims too weak to resist are enslaved, possessed, fated to do the bidding of malevolent beings. Although God's original creation of daemonic beings might have been good, said Origen, once separated from their Creator, all daemons inevitably "fall." Some fall a lot, resulting in a large ever-increasingly population of "fallen angels" ready to tempt and misguide men into committing lewd acts, requiring esoteric rites of exorcism and purification.[10]

With celibacy increasingly mandated for priests and church officials, erotic attraction was a prime demonic weapon. Solitary hermits and monastic communities withdrew from the world into the solitude of the "desert" to prove their valor in hand-to-hand battle with demonic adversaries. Locked in the solitude of their cells, monks faced down demons, bravely exposing themselves to temptation from evil thoughts and lustful "movements" in the body that defied obedience to vows of chastity. To train for final victory over demonic eros, monastic manuals outlined tactics of resistance: fasting, prayer, sleeplessness, exposure to cold, brisk walks to strengthen a monk's resolve to resist the blandishments of demons that appeared in the shape of women, or that provoked an actual woman to seduce a monk with her "beauty." Forced into contact with women, some of the more zealous made a point of smelling rotten female corpses, or mixed the sweetness of incense and perfumes used by women with foul odors. Every supposed vice had its troop of demons: idolatry, adultery, homosexuality, theft, drunkenness, but most numerous and dangerous were those devoted to eros. Bands of pagan daemons were said to "inhabit" the ruins of temples once dedicated to Rhea whose followers were "effeminate" or to Apollo whose followers supposedly "fornicated" with women and boys. Ambitious monks spent nights among toppled stones and broken statues listening for emanations, testing the power of their Christian courage.[11]

Bishop Eusebius, the biographer of Constantine, attested to the continuing menace of these haunted pagan ruins. The Emperor, reported Eusbeius, once spied out with his imperial "eagle eye" the remains of a pagan temple near Aphaca in Phoenicia.

> Like a keen-sighted eagle in his heaven-ward flight is able to see from his lofty height the most distant objects on the earth, so did he while living at the imperial palace of his own fair city discover as from a water tower a hidden and deadly snare of souls in the province of Phoenicia. It was a grove and temple, not situated in the middle of any city or public place as is usually the case for splendor of effect, but off the beaten and frequently traveled road, on the slope of the summit of Mt Lebanon and dedicated to the foul demon known by the name of Venus.

Eusebius does not explain the magic that would have allowed Constantine to see from Constantinople all the way to Mt Lebanon. One thing clear in his account, however, is the difference between the new imperial Christian cult and the rites of pagan devotion. The old sacred sites and shrines were places of retreat, meditation, spiritual insight, and celebration. A Christian church was a symbol of authority, built and placed "for splendor of effect" so as to chasten and awe believers with the power of the Lord.

Immediately in Eusebius's description of the pagan sanctuary came the charge of obscenities.

> It was a school of wickedness for all the votaries of impurity, and such as destroyed their bodies with effeminacy. Here men undeserving of the name forgot the dignity of their sex, and propitiated the demon by

> their effeminate conduct; here too unlawful commerce of women and adulterous intercourse, with other horrible and infamous practices, were perpetrated in this temple as in a place beyond the scope and restraint of law. (*Life of Constantine* III.55, 5–6)

Whether a rural brothel was run out of an abandoned temple, or the supposed obscenities were a figment of Christian imagination, there was no question in Eusbius's mind. The ruins of pagan temples crawled with demon creatures waiting to entice holy men to break their vows and lure chaste Christian women into ungodly sexual rituals. Immediately, he reported, the emperor ordered the temple destroyed so that "those who before lived without restraint, through the imperial threat of punishment, learned self control" (*Constantine* III.55, 7).

* * *

No one dwelt on the continuing threat of daemons, now the Latin "demons," as compulsively or passionately as Augustine, Bishop of Hippo in his masterwork *City of God*. Defending Christianity against the charge that disrespect for traditional forms of religious devotion brought on the barbarian invasions, Augustine condemned all pagan religion as the worship of demons, taking his clue from Platonists like his fellow North African Apuleius.[12] Augustine categorically rejected the distinction made by Roman writers like Varro and Plutarch between poetic myth and "state-civil religion." No, insisted Augustine, all pagan religion is the same, permeated through and through with worship of demons. He ridiculed Varro's sophisticated account of the origins of religion as the invention and use of symbols, rituals, and allegories that help humans to see "with the eye of the mind" (*CG* 7.5, see also 3.4). For Augustine there were no revelations to be had in any non-Christian temple or shrine, only obscene practices dedicated to demons like Aphrodite and her Roman counterparts Venus and Cybele. Could a demon-god like Aphrodite really have had sex with the man Anchises and give birth to Aeneas? After a moment of hesitation, Augustine's answer was yes. Did not fallen angels have sex with the "daughters of men" in the Old Testament? Were not the evil giants born as a result? (3.4–5). Like Philo, Augustine had no doubt that "abominations" demanded by demons were regularly celebrated in Roman temples and theaters.

One of Augustine's great disappointments was to find that many of the Platonists he admired and relied on to provide a rational basis for Christianity condoned daemonic sacrifices at pagan shrines and attended pagan festivals. In a collection of pronouncements from pagan temples on sacred matters, Porphyry, the authoritative spokesman for Plotinus's third-century consolidation of Platonic doctrine, reported pagan responses to the question—What is a man to do to recover a wife who has rejected her family and gone to join the Christians? The results were not flattering to Christians. Priests of Apollo said that a man in this predicament could do nothing, since any sect that worshiped a dead man and a criminal put to death by "right-minded judges" must have already hopelessly corrupted his wife. Only slightly more positive was an oracle of the goddess Hecate. Jesus was a good man with an immortal soul, said the priestess, but hardly a god. He caused fateful error in Christian believers who, fixated on the false divinity of a mortal man, were unable to know true divinity. "You are not then to speak evil of Jesus but to pity the folly of men," the priestess concluded, "because through him men's danger is imminent" (*CG* 19.23).

Augustine's first thought was dismissive. Who could be so foolish as not to see that such answers must have been contrived by a "clever" man who wanted to discredit Christianity and who passed on the answers to a gullible Porphyry? His second thought was demons. If these oracles of Apollo or Hecate were not stories made up by enemies of Christianity, if they were indeed pronouncements made in temples of Apollo and Hecate, they were the responses of "demons" loathsome personified spirits speaking from pagan temples and shrines, drawing gullible Christians to their obscene ceremonies. He cited Cybele's rites as particularly offensive.[13] The "demon Cybele", said Augustine, "deceived Scipio Nasica who brought her worship to Rome where shameful games are held in her honor" (*CG* 2.5). She demands loud percussive music played with drums, castanets, flutes, and cymbals. Her priests are "effeminate, unnatural, and perverse." Cybele, Aphrodite, Aphrodite's Roman equivalent Venus, as well as the goddess Caelestis whose festivals his family attended in North Africa, all were demons out to corrupt the human race.

In Diotima's teaching, priestly arts are spiritual bridges "between" human and divine; in Augustine's Christian imagination, any such non-Christian pathways are populated with demons existing between the two worlds of heaven and earth. It is the very "betweenness" of demons, warned Augustine, that makes them so dangerous in that they are both immortal and also inflamed with passion. Why, he asked, had Plato banished the poets and dramatists from his ideal republic if it was not to urge his followers "to despise the shameful commands of demons and to detest their impurity, so as to choose instead the splendor of virtue" (*CG* 8.14). A deep ambiguity lies at the heart of this Christian obsession with demons. If the gods to whom prayers went up from pagan temples were simply nothing—no more than molded bits of marble and wood or the empty imaginings of poets—there would seem to be no need for such urgency, no need for constant warnings of danger and pollution. Why should such beings have been worshiped? asks Augustine. Are they creatures better than us, just because they can fly or have immortal bodies? (*CG* 8.15) And how can demons be gods if they have passions like anger and desire for tribute. If they are gods, they would not be moved by such prayers; if they are beasts, there would be none of the rational element necessary for passions like anger and desire.

In *City of God*, Augustine returns to the subject again and again. Demons are arrogant, deceiving, they feign divinity, they persuade people to be bad, they take delight in the depiction of crimes, they teach bogus magical arts. Would not real gods know what happens on earth? Don't they know what goes on in a man's mind? So, why would a god require these repugnant creatures to bring back messages? And why would a god believe such deceiving creatures or want to associate with them? How could a god believe what a lying demon tells him? Are good demons actually Christian angels and bad demons fallen angels guilty of "transgression"? Are any demons good? After much soul-searching, Augustine decided the answer must be no. Yes, there are angels, and no matter if Platonists call them gods, they are creatures less heavenly than the true God. Only God should be worshiped, never angels. And yes, there are demons, messengers of false gods that are not the one God of the "Christian trinity." Are demons "knowing" as Socrates had claimed?[14] If so, it must be the kind of temporal knowledge of material and transitory things that any true "unfallen" Christian angel would hold cheap. Inflamed with love of the unchangeable and ineffable beauty of God, why would an angel care about material beauty? If angels have knowledge of material things, it can only be the shadow of eternal Platonic Forms of beauty, barely visible in material things, faint reflections of the one and only beauty of God (*CG* 9.22).

Disconnected phrases gleaned from translations of parts of Diotima's teaching in the *Symposium* continue to surface. Demons "carry to the gods the prayers of men and to men the responses of the gods" "for no God has intercourse with humans" (*CG* 8.20). But, Augustine struggles, what can this mean? If a human cannot speak directly to God, how much worse it is to think that a good man, "a stranger to the crimes of the magic arts, would have to use as intermediaries demons who love those crimes?" Wouldn't it be more appropriate if demons relied on and listened to good Christian men?

But never did Augustine doubt that demons exist. Demons live in the air. They like violent rites. They demand gifts. They have passions. Always evil, they are in no way "between" God and man, in the sense of being "better" than men. The true God is all-powerful. He needs no messengers, not when he can communicate directly with a chosen few. If he has a messenger, that role is reserved for Jesus. Apuleius and Platonists like Plutarch were wrong. Demons are never good, never benevolent, never revealing of any truth.

> We must believe them to be spirits, most eager to inflict harm, utterly alien from righteousness, swollen with pride, pale with envy, subtle in deceit. They dwell in the air as if in a prison keeping with their character because they were cast down from the highest heaven, and have been condemned to live in this element as just reward for irredeemable transgressions, and they prey on nonChristians, some of whom are afraid and make temples to them. (*CG* 8.22)

No one, not even Saint Paul, and not even a celibate Christian bishop like himself, is ever safe from the evil they represent.

In the place of the old daemonic spirituality, something dark and ominous had been introduced: irrational fear, hidden evil, formulas of hatred repeated by rote and periodically revived and repeated, witch hunts, the inquisition of heretics, crusades against nonbelievers, paranoid nightmares of horned and tailed devils secretly worshiped in dark Satanic rites, and, inevitably, the dangerous deformation that Plutarch called "superstition." Instead of inspiring creative activity, eros was now sexual sin and shameful lust, justifiable only in marriage and only for procreation. The old nexus between the attractions of divine beauty and happiness in mortal life was torn apart. Diotima's beauty, especially in its feminine forms, was the demonic lure that tempts believers away from God.

CHAPTER 7

Saint Augustine and Concupiscence of the Flesh

Not in rioting and drunkenness, not in chambering and impurities, not in contention and envy, but put on the Lord Jesus Christ and make not provision for the flesh in its concupiscence.

—Romans *13:13–14*

Like Phaedra in Euripides's *Hippolytus*, Augustine had passions not easily denied. His struggle to tame the bad horse of the *Phaedrus* was prolonged, painful, and never completely successful. In his *Confessions*, he traced every agonizing humiliating step and misstep of the drama in colorful self-absorbed prose. As a child he was subjected to the demanding Christianity of his mother Monica, but her faith in the love of Christ and insistence on moral rectitude had little power over his youthful impulses, especially after puberty when he experienced a strong upsurge of sexual appetite. Dutifully, he read Paul's letters advising Christians on proper sexual conduct, but Paul's praise of the superior virtue of continence was no match for an energetic young man's passion. Encouraged by his father, Augustine gave in, as he put it, to "vile lust."

A few years later, reading Cicero's introduction to Greek philosophy for Latin readers, Augustine had hope he might rein in his physical appetites with philosophic reason.[1] Impressed by the high spiritual tone of Platonism as described by Cicero, he tried "winging upwards" from earthly delights to true Wisdom with mixed success. More seductive were the teachings of the Manichean sects gaining popularity in North Africa and Rome with their promises of occult revelation, secret lore and ritual, influential membership, and convenient explanations for moral weakness. The world of the Manicheans was a battleground between an evil kingdom of dark matter and the spiritual light of divine wisdom. Here, thought Augustine, was an explanation, even an excuse, for the violent conflict he experienced between reason and the wild horse of sexual desire. No matter how his rational soul struggled to grow its wings, his material body pulled him back down into the mire of physicality. His mother, shocked at this new alliance, locked him out of the house. Eventually, under the influence of the powerful and persuasive Bishop Ambrose in Milan, he weaned himself away from Manicheism.

What kind of a man was this? To hear him tell it, as a child he was vicious, crying when hungry and fighting when he did not get his own way. The innocence of childhood was a myth, he wrote, promoted by self-deceiving parents. As a youth he engaged in petty thefts, lied, played games instead of doing his homework. At puberty he gave himself up to lust. Strongly attracted to women, he engaged freely in causal affairs, and when finally he "fell in love," he suffered painfully the humiliating jealousies and angers described by Lucretius. In the end, he compromised with "concupiscence" as advised by both Lucretius and Paul. He took a "concubine" and settled into a cohabitating partnership that lasted for seven years and produced a son.[2] He envied, he said, men like his friend Alypius who had little appetite for sex and so little need for such medicine.

> For in this [Alypius] was wonderfully chaste. Early in adolescence he had the experience of sexual intercourse, but it took no hold on him, rather he felt remorse and revulsion at it, and from then until now lived in complete continence. (*Con.* 6.12)

But on Augustine, sex did take hold, and even after his conversion and a vow of priestly celibacy, he found it impossible to completely extricate himself from the "mud and muck" of bodily desire, if not in deed in thought in the form of arousals and dreams at night he was powerless to control.

Few of Augustine's youthful delinquencies are remarkable—boyish pranks, temper tantrums, avoiding homework, liking games, sexual impulses. What is unique is the vehemence of his readiness to confess his viciousness and blame others for his faults. His teachers beat him when he did not do his work, further enraging his spirit. His teachers failed to give him true knowledge and were only interested in teaching him the skills necessary to make money and achieve high office. The curriculum he was offered, based on Greek and Roman literature taught under the guise of increasing language abilities, inflamed his passions. Love stories like Virgil's tale of romance between Dido and Aeneas aroused and distracted him. Myths of the rapes of Jupiter attributed divinity to evil acts and made it seem that committing a crime was imitating the gods. "Not any more easily are words learned for all this vileness," he concluded in heated rhetoric, "this wine of error drunk to us in words by intoxicated teachers (*Con.* I.16).

As he tells the story of his formative years, everything in his environment encouraged veniality. His mother allied with his father to refuse the early marriage that might have drained off his lustful urges. As a teacher of rhetoric, his students were rebellious, "more vile even than he was himself." In Rome, where he had hoped to find a better class of pupils, he was outraged to find that students left his classes to attend those of a cheaper teacher. Writing his *Confessions* years later, rage still rises up in him at their abandonment of him. "In truth, these are base people, and they go a-fornicating from Thee, loving those fleeting mockeries of things temporal and filthy lucre which fouls the hand that grasps it, hugging the fleeting world and despising thee" (*Con.* 5.12).

With his own treatment of others, he is more lenient. Without self-chastisement, he tells a chilling story of tricking his mother to prevent her traveling with him to Rome. Worried about the sea trip and determined to go with him, she accompanied him to the harbor. Telling her that he had to meet a friend and would have to wait and sail the next day, he convinced her to spend the night near the dock at a hostel. While she slept he slipped away, leaving her to wake up in the morning and find him gone. He described her "frantic grief" without regret. God, he says, "chastened the earthly part of her affection for me with an allotted scourge of sorrows." He was right, he said, to take no note of her anguish. In trying to thwart God's plans for him, his mother only suffered the inevitable penalty for original sin. "In her agony there appeared in her the inheritance of Eve as in sorrow she sought what in sorrow she had brought forth" (*Con.* 5.8). Birth, child-bearing, child-raising, the spark of transcendence in the teaching of Diotima, in Augustine's Christian imagination, are a source of sorrow and pain.

In Milan, his concubine was a last impediment to advancement in Christian circles. More and more his mother urged him to make a properly legal and advantageous marriage. Finally he complied: "I wooed, I was promised chiefly through my mother's pains so that once married health-giving baptism might cleanse me" (*Con.* 6.13). It was no love match. The promised bride was two years underage.

There would be a wait for the ceremony. But a condition of the engagement was that his live-in mistress be sent away. Augustine notes his own pain at separation but not that of his lover and the mother of his son.

> My concubine being torn from me as a hindrance to marriage, my heart, which clung to her, was torn and bleeding. She returned to Africa, vowing never to know any other man, and leaving with me my son by her. (*Con.* 6.15)

With her gone, he confesses, he wasted no time in procuring a temporary substitute so as to "keep up the disease of my soul," all the time suffering "great pain" in missing the one who had left. What she, the cast-off mistress, felt and suffered as she left lover and son behind as a permanent widow and perhaps an outcast can only be imagined.

Looking back on these events in the *Confessions*, Augustine sees God's steady and watchful hand engineering his eventual salvation. Although he vehemently rejected any Stoic ideal of calm in the face of adversity as impossible considering the pains and miseries of human life, his certainty of Providence provided consolation for the suffering he caused to others. If he had to leave his mother on the shore in abject despair, it was part of God's plan to bring him to Christ. If the life of his mistress was ruined, it was part of the painful but inevitable process by which he was to achieve salvation and immortality. "Thy right hand was continually ready to pluck me out of the mire," he tells God, as he confesses to sin after sin, of lust, of intellectual arrogance, of selfishness, always strengthened by the certainty of being singled out by God and predestined to share in the eternity of heaven (*Con.* 6.16).

The way to Augustine's Christian "telos" is in stark contrast to spiritual growth as described by Diotima. As a student in freewheeling Carthage, Augustine began with need and longing for the sweetness of loving and being loved, but immediately, once he acquired a desired object, the experience of love was "defiled with the filth of concupiscence." Instead of moving on to see beauty in other bodies and personalities, he was "providentially" checked by "God in his great goodness" ensuring that joy in love was immediately sprinkled with "bitterness" and he was "scourged by the iron-burning rods of jealousy, suspicion, fears, angers, and quarrels" (*Con.* 3.1). When he finally formed a bond with one woman and lived with her, it was not out of love but out of utility. In the spirit of Paul's advice to the Corinthians, cohabitation provides a steady dose of convenient temporarily monogamous physical gratification, with the child eventually conceived no more than an unwanted consequence.

> In those years I had took one woman, not in what is called lawful marriage, but one whom wandering passion and lack of judgment had brought my way. Still I had only one and remained faithful to her, and with her I experienced first hand what difference there is between the self-restraint of a marriage-contract entered into for the sake of having children, and the bargain of a lustful love, where children are born against their parents' will, although once born, they require love. (*Con.* 4.2)

In none of these experiences, as described by Augustine in retrospect, was there appreciation and erotic joy in the physical beauty of a loved one, or passionate attraction and appreciation of generative possibilities in another's beauty of soul. Nor was there any question of advance to passionate love of human institutions, customs, laws, or knowledge. In Augustine's Christian universe, erotic desire of any kind is entanglement from which a man has to free himself before he can fly up on the wings of his Platonic soul to a final reunion with God in heaven.

* * *

As the Christian Roman Empire imploded, Augustine felt ever more insistently the responsibility of making clear how heavy was the weight of "concupiscence" on the human condition. Much had changed. Pagan temples had been destroyed, their treasure confiscated and in Christian coffers. Public and private devotion to pagan deities was illegal. He was no longer a youngish man converted to Christianity telling of his difficult road to faith; he was a powerful Bishop, an authoritative figure, seasoned by years of church administration, Sunday sermons, and violent suppression of deviant sects in North Africa. In the face of a resurgence of paganism, it was his sacred duty to smooth out inconsistencies and uncertainties in Catholic doctrine and establish an orthodoxy that would survive. No, he argued in *City of God*, it was assuredly not Christian Rome's refusal to perform rituals and rites to pagan gods that brought about the sack of Rome by the Visigoths. It was not the neglect of daemonic pathways that bridge the gap between human and divine. It was Eros, chief of demons, that had to be exorcized from the human soul.

Augustine's understanding of that soul was Platonic with adjustments for Christian belief. The soul described in the *Phaedrus* and *Republic* is a battleground between reason and appetite. Once bodily appetite is under control, the rational part of the soul can begin its ascent to the heavenly forms, progressing from mathematics to geometry, to dialectical inquiry, and finally to the revelation of the supreme illuminated form of the Good. In the process, the rational soul frees itself both from the body and from the "animal soul" and prepares for death, after which it will either be reincarnated into another body or win permanent disembodied life in heaven along with the gods. This was the Platonism that both Philo and Augustine marveled so remarkably reflected Christian values of continence, virginity, and celibacy. This was the Platonism that was to replace degraded pagan eros with a purified passionate love of God that could only be Christian. As with the Stoic Logos, changes had to be made. If the Platonic One was to be the God of *Genesis* who created the world and everything in it, the bad horse erotic urges of the animal body would have to have been created by God. To think otherwise was to revert to the heresy of Manichean dualism, in which demonic forces of evil compete with God for mastery of the cosmos. This thesis, convenient as it might once have seemed as explanation and excuse for veniality, Augustine rejected. Not only did the idea of two cosmic powers contradict Christian monotheism, it allowed Manicheans to shirk responsibility for sin by blaming it on matter and the body.

If God created both body and soul, the created body could not be responsible for sin. It cannot be that "amor" is bodily and evil, and "delectio" or "caritas" is of the soul and good. In fact, insisted Augustine, in Scripture "amor," "delectio," and "caritas" are used interchangeably. Passion, neither good nor bad in itself, is what moves a man. Love—which moves a man toward something—is good or bad depending on its object, just as hate is good or bad depending on its object. Love of God is good, love of lust bad. Hatred of evil is good, hatred of good bad. The difference is not between different kinds of passion but between "what is to be loved" and "what is not to be loved" (*CG* 14.7). Passionate fear of death, hope for heaven, hatred of evil, and love for what is holy, all are good passions because all have reference to the one true and legitimate object of love that is God.

Food, music, wine, glory, fame, all, according to Augustine, are inferior objects of desire, but no vice was condemned by him so harshly and so obsessively and in such hyperbolic terms as sexual love. At puberty he was tossed into a "whirlpool of lust." "Foul yearning" bubbled up out of the muddy concupiscence of his body, drowning him in a sea of sin. His perverted will turned to base illicit pleasures; he floundered down into abysses of foul mire. Eros was loathsome weight dragging him down, chains that rivet him to earth and kept him from rising to

God, prison bars that trapped him in excrement, filth, and unclean thoughts. There was one bright note. No matter how avidly he sought after and achieved sexual release, he was left unfulfilled and unhappy. This he took as evidence of the grace of a God who providentially mixed bitterness with illicit pleasures (*Con.* 2.2). Tasting water from the poisoned well of sexual desire, he was not satisfied, but left yearning for some other eternal object of his desire, his very dissatisfaction enlivened with a strange eros of its own. In this way, reasoned Augustine, all his unclean yearnings providentially brought him closer to the one proper object of human desire, the only love object that would not disappoint, the only love object not tainted with lust. To love anything but God is idolatry. To love people or things or worldly affairs is to cut oneself off from the only salvation, which is God's grace.

> Who to tell, how to tell, how the weight of concupiscence presses us down into a steep abyss, but yet how caritas raises us up again by means of your Spirit, the Spirit that moved over the face of the waters...[Pressing us down] are affections, loves, the uncleanness of our spirits flowing down with the love of worldly concerns, but then the holiness of Your Spirit raises us up again with the love of freedom from such concerns...We are inflamed by your gift, and are carried up, we are on fire and up we go. In our heart, we ascend the ascent. We sing a song of approach, on fire with you, burning with fire of your goodness, on we go, upward we go into the peace of Jerusalem. (*Con.* 13.7, 9)

No mortal bride or bridegroom approached the bedchamber with more ardor and expectation of consummation than Augustine contemplating his final union with God.

At the same time, tragically, the very organ that in the bridal chamber ecstatically joins bridegroom and bride in nuptial bliss stood in his way. No matter how fervently his soul might desire God, to his chagrin there was a part of the human body he could not control. As Paul noted in despair, the male member did not always obey a man's will. A man might fervently will to resist physical arousal, yet his member comes erect. Equally fervently he might will to be aroused, but his member fails to respond.

> Whether they confine themselves to lawful pleasures or transgress to unlawful pleasures, those who delight in [sexual pleasure] are not moved to it by their own will. Sometimes lust importunes them in spite of themselves and sometimes it fails them when they desire to feel it, so that even though lust rages in the mind it does not stir in the body. (*CG* 14.16)

Try as he might, Augustine was powerless to heed Paul's admonishing voice: "It is best for a man not to touch a woman" (*I Corinthians* 7.1). Too young to marry, the best he could manage was to set up household with his concubine and so achieve some degree of containment of his sexual impulse. Once released from that entanglement and vowed to celibacy still he was troubled by lewd thoughts.

Eros presented both a moral and an intellectual challenge to Christian belief. If God created man in God's own image but with a sexed body, how could that body or a part of that body be evil? Why would an omnipotent God make a man so that it was impossible for him to be good? If it was not the fault of his body, then it had to be his mind. But it could not be the fault of "reason," the part of the soul Platonists like Timaeus rightly identified as made in God's image. And to blame it on weakness of will, on the failure of the good horse of the *Phaedrus* to keep the bad horse under control, only raised a further question. Had man been created with a defective will? Again, given a God who was an all-powerful Lord, Lover, and Logos, the answer had to be no.

Popular wisdom held that "free will" was the origin of sin and that men and women could achieve enlightenment and virtue by their own efforts. Given his own experience, Augustine rejected that possibility. He willed, but when he ended by doing something "against his will," it seemed that in him were two different wills, a good will and an evil will. The logic was inevitable. If God punished him with an evil will, the punishment had to have been just, but for what fault had it been administered? From where had come this "bitter herb" of eros planted in his heart? The devil? But this was to fall back into Manicheism. Somehow the irrepressible will of eros had to have come from God.

By the fourteenth book of *City of God*, Augustine was ready with a Christian account of the origins of erotic desire that would persist for centuries. In Diotima's parable, the birth of Eros is providential. In celebration of the beauty of Aphrodite, human longing and need find "means" of transcending mortality and Eros is born. In Augustine's Christian philosophy, the birth of Eros is punitive. God in his goodness created Adam and Eve and placed them in Eden, free from erotic desire and in full control of all parts of their bodies.

> Marriage worthy of the happiness of Paradise would have had desirable issue without the shame of lust had there been no sin. How that could be now there is no example to teach us. Nevertheless it ought not to

> seem incredible that one member might serve the will without lust then, since so many serve it now. (*CG* 14.23)

Without difficulty or hesitation once the decision was made to have a child, Adam's sexual parts would have accepted direction from his mind like a docile machine. He would simply "stiffen" his member as he might his arm and insert it into the proper receptacle on Eve's body, an act accomplished without desire or erotic pleasure. Then came the fatal disobedience. Eve heard and heeded the call of the serpent, disobeyed God's command not to eat of the tree of good and evil, and convinced Adam to do the same. From then on human will was tainted, a taint transmitted in male semen to all Adam's descendents. Humans were condemned to death, painful childbirth, and hard labor. Even worse was the punishment of unwilled erotic desire. In retribution for their disobedience, Adam and Eve experienced a new "disobedient motion of the flesh." Human genitals did not change in form, reasoned Augustine, but the primal pair covered them with fig leaves to show that they understood that those members were now shamefully out of their control. The soul, reveling in its own liberty, scorning to obey God, had been deprived of the command it formerly had over a part of its body. Because it willfully disobeyed its superior and Lord, it would no longer be able to master its own inferior servant flesh as it might have done had it remained subject to God (*CG* 13.13). Eros was a curse, imposed by a jealous God. Marriage and the bearing of children, which could have been pure and godly, were forever compromised. God asked little of the primal pair, said Augustine, only that they abstain from eating of the "tree of knowledge," a small thing, just enough to remind them that God was Lord and Adam was servant. The command should have been easy to obey, which proved the justice of the punishment. Augustine concluded, "Whoever thinks such punishment is either excessive or unjust shows his inability to measure the great iniquity of sinning when sin could have been easily avoided" (*CG* 14.15).

In the troubled books of *City of God*, with hope for a "city of god" on earth fading, Augustine traced the fatal disobedience of Eden back to the marital eros praised by Plutarch. Contrary to appearances, he said, it was not Eve that caused the fall. The serpent, prompted by the arch-demon Satan, approached her because that demon spirit knew that she was weak and could be deceived. The real culprit was Adam, more specifically the love Adam had for his wife. If Adam ate the apple of knowledge, it was not because he believed the apple was good or beautiful, but out of "the draw of kindred feeling" for Eve—because he could not bear "to be severed from her companionship" (*CG* 14.11). In short, he sinned out of love. In the end, Eve was less to blame. In her female weakness, she sinned unknowingly.

It remained to account for the "weakness" in Eve that made her give in to desire for what was "a delight to the eye" and "good for food" even though it was forbidden. And on the subject of whether women or men were the most prone to sexual weakness, Paul's letters gave conflicting testimony. Paul wrote the egalitarian: "In Christ, there is no Jew or Greek, no slave or freeman, no male or female" (*Gal.* 3.26–28). At the same time, to quiet controversies in Corinth, Paul used the secondary creation of Eve to mandate the subordinate role of women.

> The head of every man is Christ, the head of a woman is her husband...A man ought not to cover his head, because he is the image and glory of God, but a woman is the glory of man...For man was not made from women but woman from man. Nor was man created for woman, but woman for man." (*I Corinthians* 11.3–9)

In the *Confessions*, Augustine resorted to the body architecture dictated by Timaeus's Creator. God, reasoned Augustine, created both man and woman with a divided soul: "one part dominate because it reasons, another obeys because it is subject to that guidance." But Eve's disobedience mandated a further hierarchy. "In the same way our natural appetites need to be subject to the reasoning power of the mind in order that the actions to which they lead are inspired by principles of right conduct," woman in the sex of her body should be subject to the "sex" of her husband (*Con.* 13.32). So that the "inferior things" originating in the "women's quarters" do not distract the rational male mind in its contemplation of eternity, the female part must be subdued. If woman "symbolizes" the part of man that attends to temporal things and man the part of man that attends to intellectual godly things, women as the distracting symbol of the temporal must "cover their heads" and keep silent even though they also possess some bit of rational soul in the image of God. With this tortured logic, but in a direct line of thought from his own personal struggles with sexuality, Augustine came to the conclusion that erotic desire is punishment for an original sin of disobedience. He and every mortal man served out a life sentence of damnation, reprievable if at all by baptism, Christian ritual, and the unmerited and unpredictable gift of God's grace. No man could be moral from his own will to be good.

Every human infant was born under the "dominion of the devil," every chaste pagan virgin and pagan wife was doomed to hell.

* * *

It is a difficult moment in Christian thought. Certainly, once the hope of imminent apocalypse faded, it would not be possible to require strict celibacy of all Christians. Not only would most fail the test, if they succeeded there would be no new Christians to replace those that died. For practical reasons, the sex act could not be completely renounced. In *Genesis*, God created man, male and female. He enjoined male and female to be fruitful and multiply. Along with arms and legs, he gave them sexual "members" by which to accomplish the task. Admitted Augustine:

> We for our part do not at all doubt that to increase and multiply and replenish the earth in virtue of the blessing of God is a gift of marriage as God instituted it from the beginning before man sinned, when He created them male and female—in other words in two distinct sexes. And on this work of God his blessing was pronounced. (*CG* 14.22)

Given that no immediate end of the world was in sight, on the subject of marriage there would have to be compromise. A man or a woman should not marry out of lustful desire as Augustine had done with his concubine. But marriage contracted like his own engagement, without lust and with the expectation of having Christian children might somewhat "contain the lust of youth." And once children were conceived, the bearing and upbringing of them might cool some of the fire of "concupiscence." In this way, Christian procreation could be accomplished with a minimum of erotic desire and some approximation of the prelapsarian state of Adam and Eve. An eros-free marriage, with sex engaged in for the purposes of breeding and raising Christian offspring, could be the next best thing to the true holiness of celibacy.

Augustine was not without his critics. Prominent among them were the followers of Pelagius, who provoked from Augustine a long polemical letter "De Nuptius et Concupiscence" ("On Marriage and Lust").[3] Pelagius insisted that good pagans might go to heaven. He believed that newborn infants were innocent. He believed that marriage was a "holy state." Could no passionate coming together of a man and woman be condoned? he asked. No, answered Augustine. Marriage for reproduction is good, as long as it is not tainted by concupiscence, which is "the law of sin that dwells in our members" and "the wound the devil inflected on the human race." Married couples must always be on guard against this "bestial" appetite that amounts to "fornication against God." They have to train themselves to have intercourse without erotic desire. Never using any form of contraception, they have to engage in sex for the purposes of procreation and without "indecorous movements." If some degree of "nuptial embrace" proves necessary, even then it is sin if only a "venial" sin. The debate over the necessity of infant baptism, the salvation of good non-Christians, and the holiness of passionate married love ended in defeat for the Pelagians. Celestius, a follower and friend of Pelagius, was refused ordination. In 431 Pelagian doctrine was declared a heresy.

What then was to bring Christian husbands and wives together if it was not the erotic married love praised by Plutarch and Pelagius? What was to inspire in them the energetic striving to reproduce and care for their young described by Diotima? Was there to be no affection or emotional ties between husband and wife? Even more important, what can bind a Christian community together? If Christian spirituality is based only on the profession of faith, obedience to commandment, and required attendance at "assemblies," from where will come the passion of religious belief? Something more was needed, a new kind of Christian love with none of the dangerous physicality or daemonic urgency of erotic desire.

In the Hebrew Scriptures, one word "*ahavah*" had been used for the many different varieties of human attachment including parental affection, friendship, the love of Yahweh for his chosen people, the love Israelites owed to members of their own community, and the love of a man for his wife. Nor was Greek much better in providing a term that could distinguish good Christian love from the dangerous pagan "eros." "*Philia*," used in Platonic dialogues for bonds between men, occurred in *John*, but perhaps because the word carried residual overtones of Greek homosexuality, it was seldom used by Paul and other gospel writers. Instead, in his letters Paul had settled on the lesser used Greek verb αγαπάω, indicating the mild familial affection scorned by Plutarch as a poor substitute for passionate marital love. Here was a word without previous religious association, a term that might denote a new purified Christian love, a word that might indicate love as commanded by Jesus in sayings like "love your neighbor as yourself." Jesus could be said to agape-love his disciples and they him without any implication of homophilia. Jesus could forgive Mary the prostitute who washes his feet in the house of a Pharisee because she "loves" him, with no suggestion of lustful contact. Agape need not depend on passionate response to natural beauty. It was not "love of the world," of "the things of the world," or of "the present world" (*I John* 2:15; *II Timothy* 4:10). It was not love of honor, status, or the approval of others. It was love properly directed to the Christian God, as prescribed by gospel writers.

> Do not love the world or things in the world. If anyone loves the world, love for the Father is not in him. For all that is in the world, the desires of the flesh, and the desires of the eyes, and the pride of life is not of the Father but is of the world, and the world passes away and the lust of it but he who obeys God lives forever. (*I John* 2:15–17)

The cultivation of Christian agape required the reversal of spiritual development as taught by Diotima. In her teaching, the achievement of "*theophilia*," "friendship with divinity," comes after long apprenticeship in "the desires of the flesh," attraction to spiritual beauty, and the creating of lasting good in the form of happy households and just laws. Christian agape comes only when one renounces passionate attachment to human beings and human institutions and directs one's love to God. It is a mistake, said Diotima, to make love a god. In John's gospel, so close is the connection between God and agape that the two become one: O Θεος αγάπη εστιν ("God is love") (*I John* 4:8). God is the source of *agape*, *agape* is from God, and whoever has *agape* is God's child (*I John* 4:7).

If Christian agape is good purified love of the Christian God, its opposite is επιθυμία, desire itself, understood as appetites "of the flesh." Peter warned against false leaders who entice Christians with "licentious *desires* of the flesh" (*II Peter* 2:18). He warmed them about desires that wage war against the soul (*I Peter* 2:11). Paul reminded the Christian community at Ephesus: "We all once lived in the passions of our flesh, following the *desires* of our body and mind" (*Ephesians* 2:3). He lectured the divided community at Galatia: "Those who belong to Christ Jesus have crucified the flesh with its passions and *desires*" (*Galatians* 5:24). Desire is all that is evil, immoral, unclean, licentious, idolatrous. Desire is at the root of sorcery, hostility, jealousy, anger, drunkenness (*Gal.* 5.17–19). Desire was the Latin "*cupiditas*," greedy craving to be condemned and reviled, as apposed to "*caritas*" with its connotation of indiscriminant and disinterested charity toward others. If demonic concupiscence is a Christian invention, so was its Christian counterpart *agape*. Neither is Diotima's love of beauty promising the possibility of generative communion; neither is Plutarch's marital grace. The end of both concupiscence and agape is solitary, in the one case a moment of selfish pleasure, in the other individual salvation. If concupiscence is the beaten-down and tortured desire of the *Phaedrus*'s wild horse of bodily love, agape is what results when erotic response is cut off by prohibition, leaving only the "vague and powerless affection" that Plutarch predicted results when marital love is separated from sexual attraction and fulfillment.

Augustine's struggle to resist the attraction of earthly beauty was prolonged and painful. He freely confessed his susceptibility to sensual pleasure. "There is an appeal to the eye in beautiful things, in gold and silver and in all such. The sense of touch has its own powerful pleasures, and the other senses find qualities in things suited to them." He acknowledged the appeal of ambition to which he was hardly immune: "Worldly success has its glory, and the power to command and to overcome, from which springs the thirst for revenge." He praised the beauties of nature, "the beauty and glory of the stars in the heavens, and of earth and sea with their oncoming of new life to replace the generations that pass." But always would return the fear of demonic influence. Were these not "the false show or shadow of beauty by which sin tempts us?" No matter its attraction the beauty of this world is a dangerous snare, luring us to evil deeds (*Con.* 2.5–6).

> If we ask why this or that evil act was done, it is normal to assume that it could not have been done except through the desire of gaining or the fear of losing some one of these lower goods. For they have their own charm and their own beauty, though compared with the beauty of heaven they are poor and mean enough. (2.5)

Augustine's courtship of the God who ruled that heaven took on all the borrowed passion of erotic love, expressed in sensual images as if yearning for the taste, breath, scent of an absent lover.

> I came to love you, Beauty, so ancient and yet so new, I came to love you.
> Behold you were in me, but I was away and I looked for you there,
> To the shapely things that you made I shamelessly rushed.
> You were there with me, but I was not with you.
> Those things, which without you would not be at all,
> Held me far off from you.
> You called and cried to me, broke into my deafness,
> You flashed, you beamed, and you chased away my blindness.
> You exuded the fragrance of your breath, and I panted for you.
> I tasted you, and I hunger and thirst for you.
> You touched me, and I burn in your peace. (*Con.* 10.27)

Continence itself was eroticized in a masochistic ecstasy of "do to me what you will."

> Oh Love, always ardent and never extinguished,
> Dear one, my god, set me on fire.
> You order continence, give what you order,
> Order what you wish. (*Con.* 10.29)

It was dangerous ground. Always there was the chance that the pull of sensuous beauty as celebrated in rites to Inanna or Aphrodite might break loose in literal sexual abandonment. Origen, the Christian ascetic, warned against reading the *Song of Songs*. Only those who had succeeded in "crucifying" the desires of the flesh could appreciate the "beauty" of bride and bridegroom celebrated not as ordinary beauty but as the spiritual beauty of a redeemed Israel and the goodness of God. Only someone who had first learned to practice continence as preached in *Proverbs* and gone on from there to *Ecclesiastics* to develop contempt for "the things of this world" could understand the *Song of Songs* properly as praise of the heavenly beauty of God rather than the fleshy beauty of a human lover.

The need for constant vigilance was eloquently described by Augustine's contemporary Jerome in his doctrinal "Letter to Eustochium." Beware, he wrote the young daughter of his supporter and patron Paula, beware demonic "principalities" and "world rulers of darkness." Beware the "spiritual hosts of wickedness in heavenly places," beware "the demon and his attacks at noonday." Your adversary, wrote Jerome, "is the devil who goes about like a roaring lion." No one, he reminded her, no one is exempt, not when even the apostle Paul confessed to suffering "pricks of the flesh," not when even Paul had "another law in his members" tormenting him after he had fasted, gone naked, imprisoned, and whipped himself. When "lust tickles the sense and the soft fire of sensual pleasure casts over us its pleasing glow," she should pray. Avoid wine that "kindles the fire of sensual pleasure," a tool of demon lust waiting to snare young girls and lead them into sin. Struggle against the sensuous beauty of color. Do not emulate women who go through the streets with free-flowing hair, who wear colorful robes and lilac mantels, and walk with an easy stride (13). Reject dresses dyed purple and the glitter of gold threading (32). Do not "make yourself gay" with music or poetry or attend banquets serving delicacies of food and wine (29). To be loved by God the King of heaven it is better to be a woman "pale and thin with fasting" (17). Remember the punishment of the Chaldean woman in *Isaiah*, remember her punished, stripped bare, abused, stoned in public. "Do not let Zion, the faithful city become a harlot, do not let it be that where the Trinity was entertained, demons shall dance and owls make their nests and jackals build."

To sweeten the appeal of chastity, Jerome quoted liberally from the reinterpreted *Song of Songs*. "Forget your own people and your father's house, and the King shall desire your beauty." Her soul would leave its "native" land and the "demons" that abide there; it would forget and abandon worldly things. She may be naturally "black" like the bride in the *Song of Songs* because she is "born of the devil," but if she waits for the heavenly King, she can become white (25).

> Let the bridegroom sport with you within your chamber. When sleep overtakes you, he will come behind the wall, thrust his hand through the opening and caress your belly... you still start up trembling all over and you will cry, "I am wounded with love." (25)

Surrender to God, he told her and other young girls, promises sacred bliss beyond any happiness to be found in mortal love. Turn away from demon eros, wait for the divine lover who comes to you at night in spiritual ecstasy. Compare the horror of earthly marriage that promises only "pregnancy, crying babies, and household management," "fancied blessings" quickly cut off by death. For weak souls, marriage may be an "honorable state," but not for someone in love with the beauty of a God for whom the Diotimean beauties of the natural world are as nothing.[4]

CHAPTER 8

The Eclipse of Beauty

> *What we know to be most proper to God—his self revelation in history and in the Incarnation—must now become for us the very apex and archetype of beauty in the world, whether men see it or not... Christ becomes the measure, both in judgment and in redemption of all other religious forms in mankind.*
>
> —*Hans Balthasar,* Seeing the Form *69, 171*

Diotimean "Beauty" had begun to fade from "Platonic" consciousness as early as the *Phaedrus.* Introducing Phaedrus to the divine mysteries of a Pythagorean heaven, Socrates included Beauty along with Wisdom and Goodness as heavenly, colorless, formless, and visible only to the mind. Later, when the winged souls of his lovers actually take flight from earth and make their rounds in the train of the Gods, Beauty has disappeared. The liberated souls see only icons of Justice, Moderation, and Knowledge. When the lovers return to earth, beauty reappears, but both its reality and its divinity are compromised. Now it is no more than the dimly reflected glimmer of remembered Forms of Justice, Moderation, and other virtues. The closest thing to true beauty, speculates Socrates, is light, dim light on earth that lets philosophers see shadows of justice and goodness, bright light in heaven that lets ascendant souls to see what Justice and Goodness really are.

Justice and Moderation, viewed by the "eye of the mind," are philosophical abstractions, reified ideal concepts set in fixed definitions as what is truly and always just or moderate. Beauty, less amenable to dialectical understanding, is left behind in the changing visible world of the senses, not definable in the same sense as justice and moderation. This is made clear in the *Greater Hippias.* The sophist Hippias's boast that he makes large amounts of money giving speeches to young men explaining to them the beautiful ways of life they should pursue gets a typical Socratic response. This must mean that Hippias knows what beauty is. But when Socrates asks that Hippias share that knowledge, all Hippias can do is provide examples of beautiful things. Even after Socrates takes the lead and the two settle down to debate various definitions—what is fitting, useful, gives sensual pleasures—all attempts fail, resulting either in paradox or contradicted by counterexamples. Socrates and his friends, says the deflated Hippias, do too much playing around with words. Left by himself, Socrates laments that once again he must go home unenlightened.

Diotima's beauty itself continued to cause problems for later Platonists. On the one hand, it could seem that the *Symposium* should serve as a foundational text for Platonism. Was not beauty itself one of the first references to ideal entities existing apart from the physical world? Did not the *Symposium* along with the *Phaedrus* mark the line where the skeptical and agnostic Socrates disappears and Platonic doctrine begins to emerge? For Plotinus, founding a Platonic school in third-century Rome, the idea of a hierarchy of abstract and ideal objects of philosophical knowledge apart from the physical world was at the core of that doctrine. Certainly in the third century there were reasons to want to escape the physical world. Rome struggled to hold on to an overextended empire, fighting enemies both at home and abroad. Apocalyptic sects—Christians, Manicheans, Gnostics, Zoroastrians—shouted alarm, proclaiming that a final cosmic catastrophe was at hand. Disease was rampant—typhoid, diphtheria, plague, blindness, venereal disease—and Plotinus himself suffered from many of these ills. "Plotinus, the philosopher of our times," wrote his student and follower Porphyry in his *Life of Plotinus*, "seemed ashamed of being in the body" (1).

The message of Platonism, as Plotinus taught it, was simple. The key to the good and happy life in a hopelessly flawed world cannot be pleasure, as Epicureans would have it, or living according to "nature" as taught by the Stoics. It was separation from the body for the life of a rational soul in contact with the "One" absolute Good. To achieve that state meant putting aside ambition for political power or civic office, as well as cultivating indifference to one's own and others' misfortunes.

> Suppose death strikes at his household or at his friends; he knows what death is, as do the victims, if they are among the wise, know also. And if death taking from him his familiars and intimates does bring grief, it is not to him, not to the true man but to that in him which stands apart from the Supreme, to that lower man in whose distress he takes no part. (*Enneads* 1.4.4)

Plotinus's Platonist achieves happiness by ignoring the body and the animal part of his soul except for what is necessary for physical survival. He lives the "life of the mind," engaging in abstract thought and advancing on to contemplate the form of a unitary Good. Gone is the long "work" of any praxis of love. Gone is the need for a variety of daemonic "means of access" to the divine. A Platonist ignores visible beauty. He gives up the attempt to create beautiful forms of life so as to move as directly as possible to the source of all beauty and goodness.

In the notes compiled by Porphyry from Plotinus's lectures as "On Beauty," Plotinus struggled to accommodate the divine Beauty of the *Symposium*. Is there a form of Beauty like the form of Justice? Or is beauty too much a part of the physical world to admit of such abstraction? Can Beauty be reduced to the intellectual judgment that something is good and therefore attractive? Or if beauty of soul is understood as an amalgam of virtues, what part does the beauty play? The common sense view is that beauty is a sensual quality. Music, color, sculpted forms are beautiful. We see and hear beauty with the physical senses, not with the "mind's eye" but as proportion of forms and color harmonies whether in the natural world or in art. And how can a single thing like a Platonic form be beautiful in the same way as the ways of life, discourses, and forms of knowledge cited by Diotima are beautiful? (*En.* I.6.1; cf. *Sym.* 210c) Plotinus pondered: if well proportioned simply means a harmony of parts, a body of knowledge could be well proportioned and at the same time evil and false.

Plotinus repeated questions Socrates had been able to answer only after Diotima changed the subject to the good. Why do we respond to and welcome beauty? Why do we value it so? Why is it that confronted by what is "ugly" a person shrinks back, alienated and anxious to escape, whereas when she sees beauty she is energized and joyful? Given Platonic doctrine, it must be because the soul "recognizes" some vestige of a form dimly remembered from a previous disembodied life, arousing a purified Platonic eros that is consummated when it sees Beauty with no help from the senses. As such, the final vision of divine Beauty is for the privileged few "lovers of knowledge" able to see with the "eye of the mind." Many will be blind to the beauty of abstract justice or courage. Only a few will grow the soul wings of the *Phaedrus*: Although the ordinary person might have some idea of virtue, the vision of divine beauty is only possible for the philosopher who falls in love with Goodness itself, not for the ordinary person in love with the beauties of physical life.

But still Plotinus was not satisfied. The questions raised by Diotima's lessons in love persisted. How could one "fall in love" with full erotic effect with ways of life, discourses, or knowledge if their beauty is only in their "form"? And why does a person recoil so viscerally and bodily from ugly souls that are dissolute, lustful, fearful, jealous, in love with impure pleasures. Would it not be better to do away with eros all together? Plotinus recalled the manic imagery of the *Phaedrus*, the tormented soul dragged down into the mire and the mud, copulating with the body and sensation. No, what is beautiful must be totally separate from the body, it must be mind alone concentrated on what is "above." But again came the puzzle. What is the relation between "*to kalóv*" and "*to agathóv*," the beautiful and the good? And what is the relation of the two of them to the "One" true Good?

Plotinus began again. For a soul to become beautiful and good is for it to become like God, for from God must come both beauty and living beings. Beauty then comes first, beauty that is also the good. Then must come Intellect, also beautiful, and then the soul that is beautiful because of intellect, and then everything else including ways of life and knowledge that are beautiful by way of the soul (*En.* I.6.6). But no. For a Platonist, the primal "One" must come first, then Intellect, and from Intellect come souls. From souls come knowledge and the arts. Beauty? Perhaps in the end it is expendable, no more than a deceptive shimmer glossed over Unity of Form, an insignificant byproduct of the Platonic One, the only true object of Platonic desire.

> If anyone sees it what erotic passion will he feel, what longing will there be in his desire to be unified with it, what a shock of delight! The man who has not seen it may desire it as good, but he who has seen it glories in its beauty and is full of wonder and delight, suffering a shock which has no hurt in it, passionate with true eros, and piercing longing. He laughs at all other eros and despises what he thought to be beautiful before. (*En.* 1.6.6)

Plotinus repeats Diotima's closing rhetorical question to Socrates with subtle differences in wording: "What then are we to think, if anyone gazes on beauty-itself, by-itself, uncontaminated by flesh or body, not on earth or in the sky, but in itself pure?" Diotima's divine beauty-itself is "not" the earth, not the sky, not any of the primal deities of Hesiod's mythic *Theogony*. In contrast, Plotinus's divinity is not "on earth" or "in the sky," but elsewhere in a world that does not figure in Diotima's integrated cosmos. The "observer" of Platonic mysteries "laughs" at earthly beauties; he lets the material world fade away, unimportant in the light of eternity. For eternity, said Plotinus, a man must give up the beauty of colors, bodies, and lovers, the beauty of rule and kingship, the beauty of laws and bodies of knowledge,

How is a soul to get to this final Platonic revelation and see what is in the 'holy sanctuary'?" Plotinus's directions for an aspiring soul contravene Diotima's lessons in love. First, one must leave bodily sensation behind. Do not run after shadows of bodily beauty. Do not sink down into the mud of lust. Be the brave Odysseus, leave behind Circe's enchanted island, abandon the seductive nymph Calypso. Set sail for the "fatherland," not by literally setting sail and going anywhere or seeing anything but by closing your eyes and entering the world of thought. At first, you may suffer only "dazzlement" or, more likely, blankness. You may have to return to "ways of life" and "works" and take a slower uncertain way up the ascent. But as soon as you begin to grasp outlines of form in real things, close your eyes and begin to mold yourself to match the form. Chisel away at what in you does not match the ideal, ruthlessly cut out all excess, until you blissfully contemplate your own perfection (*En.* I.6.8).

In "On the Dialectic" (*En.* 1.3), Plotinus provided some alternate routes. A "musician" might begin with the beauty of sound and work to see the mathematical and intelligible harmony behind the music so as to learn to love, not the music, but the form inherent in the sound, all of which will take time and may not be accomplished at all. The "lover" of boys of the *Phaedrus* comes closer. He responds to the visible beauty of young male bodies, closer to "form" than sound, but he will still have to go through a Platonic initiation, moving from bodies to souls, to ways of life, to ways of life and bodies of knowledge, before he can leave earth behind for the heavenly vision. A man with an innate feeling for virtue can take a somewhat more direct approach. Skipping over love of any bodily or mental beauty and beginning from the contemplation of the idea of arête or excellence, he may progress to the intelligible world of Intellect. But for Plotinus, as for other Platonists, the privileged and surest route to enlightenment is the highroad of Parmenides, reserved for the "philosopher" with the winged soul. Bypassing the towns of ordinary men and women, naturally drawn to the abstract study of mathematics and logic, unmoved by any mating of male with female, the philosopher need only perfect his own excellence. Forgetting the sensuous beauty of sound or color, he engages in dialectical inquiry, dividing and ordering the intelligible world under the one supreme form of the Good. In the end, he has no need for beauty at all. It was no more than a "screen," an alluring veil behind which true Goodness hides, a come-on unnecessary in the greater illumination of metaphysical certainty.

* * *

Beams of Diotimean beauty continued to shine out periodically from behind the blank façade of Platonic Being. In the mystical Christian neoplatonism of the fifth-century Pseudo-Dionysius the Areopagite, the first "divine name" of God is the Platonic Good that lights up an upper world of mystical revelation. But another divine name of God, said Dionysius, is "Beauty," which he evoked in Diotimean terms as the motive force behind all motion, change, and combination. Beauty "calls," said Dionysius, citing the similarity of καλός to the Greek verb καλέω—"to call." Beauty attracts and moves all things, both as the beauty of particular people and things and as beauty-itself, which Dionysius describes in terms taken directly from Diotima in the *Symposium*.

> Beautiful, as at once beautiful and super-beautiful, and always being under the same conditions and in the same manner beautiful, and neither coming into being nor perishing, neither waxing nor waning; neither in this beautiful, nor in that ugly, nor at one time beautiful, and at another not; nor in relation to one thing beautiful, and in relation to another ugly, nor here, and not there, as being beautiful to some, and not beautiful to others. ("Divine Names" IV.41)

By way of beauty-itself, he says, come "the adaptations of all things, and friendships, and inter-communions."

From the standpoint of Christian orthodoxy, this was again dangerous ground. Does including the seductive pagan "*to kalov*" warned against by Augustine, Origen, Jerome, and other Christian writers go "against the oracles," against the revelation of Christian truth in the holy Scriptures? Does not talk of eros run the risk of abandoning the disinterested charity of Christian *agape* for pagan concupiscence? Critics had to look at the meaning of "eros" as he uses it, said Dionysius in his own defense. Christian eros is not the old degraded sensual eros; it is a Platonic eros of the soul, rising above animal appetite to pure and sanctified contemplation of the one Christian God. Like Philo, he found confirmation in biblical writings.

> But that we may not seem, in speaking thus, to be pushing aside the Divine Oracles, let those who libel the Name of Love hear those

> oracles. "Be in love with It," they say, "and It will keep thee—Rejoice over It, and It will exalt thee—Honor It, in order that It may encompass thee,"—and whatever else is sung respecting Love, in the Word of God. ("Divine Names" IX.47)

There would be no such suspect reversion to pagan eros for Thomas Aquinas. In place of the dangerous "call" of beauty was now a more decorous Latin "*pulchritudo*," promising a quieter "*delectatio*," either mildly active and recreational as in play and listening to music or, in a more elevated Platonic sense, cognitive pleasure in apprehending proportion and harmony in God's creation. Aquinas's divine Creator stands resolutely alone. He copies no preexisting forms of mathematical reason, nor does he work with any preexisting beauties of nature as did Timaeus's demiourgos. As Being itself, he is Reason and Goodness. Anything good or true is true and good only by participating in that one Goodness and Truth.

Again, beauty presented problems. It was hard to ignore the passionate ecstasy of mystics like Dionysius in love with the beauty of God, but somehow any Christian beauty had to be securely contained within the one Goodness and Logos of the Christian God. Judging from the few times Thomas mentioned Beauty as one of the "transcendental" determinates of God along with Truth and Goodness, and the ambiguity of those few references, it could not have been an easy task. Beauty by any ordinary understanding is both "visible" and desirable and, as such, an aspect of the natural world. But the divine Beauty of the Christian God could not be visible or desirable in the same way a beautiful body is desirable. To see God's beauty, a Christian had to look past sensual visible appearance, bypass erotic attraction, see through existing persons or things to something that could be identified as the one Truth and Goodness of God. Then, seeing the beauty of that Truth and Goodness in the skeletal form of an original creation, he would experience not pagan eros but pleasure of a suitably disinterested and cognitive kind. "Beauty," said Thomas, "consists in due proportion" (*Summa Theologica* I.5). Proportion, in turn, he defined in Platonic terms as quantitative in the mathematical spirit of the *Timaeus*, and qualitative in philosophical terms as logical relations between genus and species, trees of dialectically defined concepts, and hierarchies of created being. Seeing the outlines of this created "formal" beauty in physical things inspires not eros that draws a believer to merge and participate, but a distanced "aesthetic" pleasure free from worldly or sensuous entanglement. True beauty, said Thomas, has to do with a kind of "knowledge" of Christian doctrine that is not "thought," "believed," or "understood", but "viewed" through a screen of physical appearance. Jesus and his deeds are beautiful not because a man and his teaching were beautiful but because a believer sees in Jesus an image of the son of God come to save the world. The human body is beautiful not because it draws a lover into intimate generative relation but because a Christian sees that it was constructed according to the specifications of the divine Creator.

If Timaeus wrote the tortured and conflicted chariot driver of the *Phaedrus* back onto the creation of a man meant to be from the beginning what he had to become, Thomas's description of the Christian God's beautiful plan for the human body replicated that model for Christian believers. At creation, said Thomas, higher rational powers were placed in the head. Appetitive powers like sexual desire were placed lower down. From there they could rebel against the will of the rational soul and come to dominate the soul, but also be controlled by the force of a Christian mind above in the head, barring animal spirits from action. The "beauty" of this "proportion" in the human body is furthered, said Thomas, by the upright design of the body, with head placed so as to view objects in the physical world not for immediate use or accessibility but so as to see beyond them to the intelligible form that bears the plan of God's Creation.

> Man has his face on top, so that through the senses, and particularly the sense of sight, which is more discriminating than the others and shows him more differences between things, he is free to be aware of objects all around, on earth and in the heavens, and from all of them gather truth that is intelligible. (*Summa Theologica* 1.91)

"*Claritas*," Thomas's name for the flashing illuminations of Dionysus's divine beauty, was harder to accommodate. Color and brightness it would seem are necessary for vision of any kind, but the divine color and brightness of created forms could not be the enticingly brilliant purple of a woman's robe or the seductive sparkle of a rosy complexion warned against by Jerome. Instead, Thomas identified *claritas* as the emanating light of the heavenly bodies, a penetrating "light of reason" illuminating icons of saints and carvings of Jesus on the Cross. *Claritas* was the Platonic sun, lighting up Forms in an "upper" world of meaning freed from the obscurity of politics and human involvement, beams of which brightened objects of Christian devotion in the form of church art. If Plato's philosopher kings in the *Republic* returned to the city to draft wise legislation, Thomas's monks and pilgrims had no such need. Christian beauty was its own reward, the glare of a vision appreciated in full splendor only when blind to the sensuous beauty of the physical world.

* * *

This partial and tentative cooption of beauty itself as transcendent aspect of the Christian God is brought to completion in the contemporary Catholic "theological aesthetics" of Hans Urs von Balthasar's *The Glory of the Lord.* Again the "call" of beauty is seen as too important to be simply ignored and left outside the circle of Christine belief. Not only does the allure of beauty present an ever-present danger of apostasy in the form of rival religious practices, without beauty Christian doctrine is itself in danger of becoming empty dogma, becoming, as Balthasar put it, no more than "syllogisms" that "dutifully clatter away like rotary presses or computers" (*Glory of the Lord* I; *Seeing the Form* 19). Balthasar approved Protestant theologian Gerhard Nebel's attempt in *The Event of the Beautiful* to infuse Christian belief with daimonic spirituality.[1] He agreed with Nebel that it is not enough to distinguish Christian theology, which depends on faith in Scripture, from Platonic philosophy, which depends on reason. It is not enough to posit a duality of inner soul and exterior God. We must make room, said Balthasar, "for the daimon that irrupts in the event of the beautiful." The daimon of the beautiful must be brought into relation with the triune God; it must be given place within the Bible" (Nebel, *Event* 148; Balthasar, *Seeing* 62).[2]

> Reality cannot be exhausted either mythically or Biblically merely with the double affirmation of man's interior space and of the exteriority of the creating and redeeming God. The phenomena compel us to establish the existence of an intermediate realm proper to the *daimon* of the beautiful, a realm which for centuries now has been progressively restricted, suppressed, and finally even denied. The history of the modern world lies just as much in the ruination and abolition of the daimonic *intermunddium* as it does in the de-Christianization of public life and culture. (*Seeing* 62)

It is the dilemma identified by Plutarch in his dialogues on religion. If the one Good or the one God is utterly removed from the natural world, it would seem that contact with divinity can come only after death. In life, soul and god are separated by an unbridgeable gulf. As a result, there is no role or meaning for the priestly arts identified by Diotima as inspired by spiritual knowledge, leaving only "mechanical" skills of craft and manufacture (*Sym.* 203a7–8). Balthasar saw just such an outcome in modern "de-Christianized" materialist utilitarian secular society. Something had been lost that the pagan world recognized as essential to human experience. Although, as Balthasar put it, there is always the danger that any pagan beauty comes from the "prince of this world" rather than from the true God, he approved Nebel's finding of intimations of divinity in Greek myth, art, temples, and tragedy. Apollo could be seen as a "mask" of Yahweh; tragic drama could be understood as reflecting a sense of the guilt of original sin. Nebel did not, said Balthasar, make the mistake of automatically identifying "daimonic" phenomena with satanic demons. He understood that the Being and Goodness of the Christian God could also be "disclosed" in beauty, even if that beauty was only visible to those with faith, even though Christian beauty is nothing unless "forged into an historical fact," and even though the daemonic might always be the demonic in disguise.

Along with Nebel, Balthasar described the transcendence of pagan beauty in terms reminiscent of Diotima's final vision of beauty-itself. "By its very nature and purpose, every beautiful thing points away from itself to some other beauty, since if we remain too long with one thing, its depth soon vanishes" (*Seeing* 65, quoting *Event* 171). Nebel, said Balthasar, saw that pagan beauty escaped the rigidity of canonical definition and the specificity of historical fact; he noted beauty's independence from Platonic hierarchies and ordered oppositions. He saw beauty's resistance to the sovereignty of mind over body, male over female, light over dark.

> The beautiful knows nothing of facts, it knows only polarity and balance, it cannot therefore acknowledge night's disengage in creation. The *daimon* of the beautiful is intent on bringing night and women into a position which creation has denied them. The beautiful is a celebration of the night and it attempts to exalt that which is weaker by the standards of creation. (*Seeing* 66; *Event* 304–309)

Nebel also saw the limitations in pagan beauty in myth and pre-Christian religion. He saw that if that beauty had value it was because it "points to" the unique event of Christian belief, the birth and death of Jesus. He saw that once that historical event occurred, the flame of pre-Christian beauty was "extinguished so that the terrible and tremendous can occur" from then on existing only in "the realm of intoxication and dreams."

But, said Balthasar, what the Protestant Nebel did not grasp was the full extent to which Christian beauty eclipsed pagan beauty. He did not see, said Balthasar, how, with the coming of Christ, beauty hardened into "diamond-hard historical fact" becoming the one "unsurpassable archetype" of all beauty, totally obliterating any previous tentative approaches to divinity. If beauty could not be left outside Christian theology, nor could it retain the glowing embers of any intoxicated or dreaming half-life of female deities and nighttime spirits. In the consuming fire of Catholic beauty, all such images were burnt away, taken up into the one Truth of Christian doctrine and the one Goodness of Christian teaching. From then on for all time, the call of beauty-itself is the call of the Christian God, a "transcendental" aspect of that God's nature along with Truth and Goodness.

Balthasar acknowledged the weakness of passionless agape. Why would a Christian love his neighbor? Why not embrace evil for his own advantage? What is to make Christian virtue attractive?

> In a world without beauty—even if people cannot dispense with the word and constantly have it on the tip of their tongues in order to abuse it—in a world which is perhaps not wholly without beauty, but which can no longer see it or reckon with it, in such a world the good also loses its attractiveness, the self-evidence of why it must be carried out. (*Seeing* 19)

At the same time, Christian love of beauty had to forever withdraw from the tainted beauty of persons, techniques or works of art, nature, forms of human knowledge or society, as well as from all other religious traditions. Once the tremendous event of the crucifixion occurred, there was only one true Beauty-by-definition: the beauty of biblical narratives, Christ on the Cross, the resurrection, the Trinity, and the "forms" of Catholic sacraments, appreciated not in the sense of profane aestheticism as the enjoyment of colors or form for their own sake but as identical with the one Christian Truth and Goodness.

> What we are used to calling "aesthetic" is as tinged with the vanity and unreality of original sin as is (enlightened) reason. Only Christ and God's Word in him in the form of suffering (the hiddenness *sub-contrario*), the historical word of Scripture reveals anew God's glory. (*Seeing* 81–82)

Balthasar acknowledged the power of beauty to move and motivate. He saw that intellectual belief in the truth of doctrine is not enough to give meaning to life. He saw that a list of commandments cannot ensure virtue. Without beauty and without the love of beauty, there is nothing to "pull" a believer in, to attract him or her to Christian truth or goodness and keep a Christian true to the canonical "Word of God."

But in Balthasar's "theological aesthetics," the enlivening moving beauty of Diotima turns to stone. When Beauty becomes "solid diamond-hard historical fact," and the event of Jesus's crucifixion is the only "unsurpassable" beauty and the archetype of all beauty, beauty becomes a thing, a set of beliefs, a code of morality, one historical narrative, one form of liturgy. Beauty is Jesus as the "suffering fool," doctrinal forms of biblical history, a celibate masculine priesthood, the Catholic sacrament of marriage, the damnation of infants not baptized in the faith. Beauty is a name, mandated, predefined as a necessary complement to the Truth of a specific doctrine and the Good of a specific tradition of religious teaching. Any ugliness in the application of that doctrine or practice of that good—death for heretics, holy wars, tolerance for pederastic abuse, ban on contraception, stigmatization of homosexuality—disappears behind a heavy veil of institutional fiat. Human history with all its beauties and ugliness disappears in the face of one narrative, one predetermined telos, one story with one ending policed for orthodoxy. Any "light" not from that God then "can be attributed to the sinner's perverted heart, his pride and his lust," and "the crowning luminosity" experienced with other forms "derives from ungodly spirit of this world and of the underworld" (*Seeing* 169).

What must be escaped, Balthasar explains in *Engagement with God*, is the "sacrality of the cosmos" (16). A Christian understands that he lives in "a world of murder and senselessness" (7). He understands that only divine commandments can redeem an "order of nature, built to a considerable extent on the principle of aggression." He sees that given the mortal world's essential "fallenness," there is no hope for "the historical future" (58). "The will to power, the principle of aggression, and the submersion of the noblest in life by the common and vulgar constantly make all human striving seem ambiguous and in the end to be vain" (59). Masters oppress workers as described by Marx, human psyches are tormented by inner conflict as understood by Freud, action is driven by the will to power as pronounced by Nietzsche, but a Christian sees that any political, psychoanalytic, or artistic hope held out by these thinkers is in vain. He sees that nature is the result of evolution driven by the survival of the fittest organisms to prey on the weak. He sees that all men are equally fallen, equally degraded, equally in need of Christian "liberation."

Most important, he sees that no "liberation" comes from mortal eros. To say in the face of death that something can come of human love—or that after the death of a loved one natural processes heal a broken heart and allow a mourner to recover and love again—"only thrusts man down once more into the domain of nature" (66). Any supposed "civilizing and humanizing of the world" is inevitably unsatisfying, "a great funeral pyre," in the words of Teilhard de Chardin, "preparation for a holocaust" (70). This, said Balthasar, was the great mistake of Old Testament Judaism: to think it could take law into its own human hands and establish an earthly kingdom of God. Instead, Christian "liberation" is based on the understanding that social and economic structures cannot be made just or harmonious.

> To think that a Christian, by his efforts, is able so radically to change the structures is mere chiliastic fantasy or the hope of an unrealistic enthusiast: The structures belong inseparably to our mortal existence, an existence that is constantly threatened because of guilt and has indeed fallen prey to sin's temptations. (*Engagement* 86)

If there is a Christian mission to "save the world," it is only to bring fellow humans joy of a different kind. Contemplating the suffering of Jesus on the Cross as the "apex and archetype of all beauty," accepting death—one's own as well as death itself—as a freely given gift from a loving God, Christians share in God's freedom by accepting the fate God has ordained for them. They participate in the sacraments of the church and do the missionary work of bringing the "joy" of those sacraments to others, especially those close to death. Blinded by the light-struck "splendor" of the crucifixion, a Christian lets the miseries of the world fade away. He embraces not the inconceivable freedom of joyful creative labor or the impossible achievement of self-understanding, or the genius of artistic creation, but the "absolute and divine freedom" of death.

Balthasar's Christian praxis turns resolutely away from anything resembling Diotima's work of love. Instead, a Christian learns to relate to his "neighbors" in a new way. He does not look at them or see them with his "natural eyes." He does not attempt to understand them using any "categories of everyday psychology." He looks at them with "eyes of faith" (51). He does not see them as oppressed or exploited—as employer or worker, husband or wife, white or black. He does not see emerging buds of beauty ready to open in their individual hearts. He sees them as possible converts. Seeing them in that way, he proceeds with the Christian mission of liberation by revealing to them their "Christian hearts," demonstrating to them the truth of Christian theology, and pointing them to the consolation of the "Man on the Cross" (55). Just as Empedocles evoked a time of strife and war, loud with the din of slaughter, horrific images surface in Balthasar's writing of the concentration camps, gas chambers, and torture chambers of World War II. But to embrace death and torture as a gift from God, to see the only real beauty in life as the prospect of death, is to sink into a despair that the intoxicated Empedocles, dreaming of future incarnations and remembering the gentler reign of Aphrodite, could hardly have fathomed.

PART III

Lessons Regained

CHAPTER 9

Religion without God

In this view of divine agency, it is incoherent to think of God working in the world apart from secondary causes, or beside them, or in addition to them, or complementary to them, or even in competition with them... Rather the mystery of the living God acts in and through the creative acts of finite agents which have genuine causal efficacy in their own right.

—*Elizabeth Johnson,* Quest for the Living God *193*

At a certain point in Plato's *Timaeus*, Timaeus breaks off his story of the formation of the world according to arithmetic, geometric, and logical principles. There was something missing from his account, he says, something that has to be put "along side" works of "intelligence," something that has to be there in the beginning before any divine Craftsman could begin his work. He will have to start again, he announces, and try to explain where the physical elements—the fire, earth, air, and water that are the material for creation—come from and what characteristics they had before the "Creator" brings order to the cosmos. Timaeus's Creator is good because he is a mind, "copying" eternally preexisting, true, and beautiful mathematical forms, but in the amended account of creation that follows, he also has to work with the preexisting "beautiful" substances that nature itself provides.

> We must now understand what are the most beautiful substances, different from each other, but able to be generated out of each other by dissolution, because if we succeed we will have the truth of earth and fire and all the substances between them. For to no one would we concede that there is anything more beautiful than these different substances, so that we must urgently try to fit together these substances that are the most beautiful, and adequately grasp their nature. (*Timaeus* 53e)

In short, there has to be a "third" thing in addition to the two Timaeus has already described, something that is not the mathematical forms and patterns copied by the master geometer and not the replicas of those forms and patterns he creates, but a precreation that goes on before any mathematical mind imposes order on the physical world.

Timaeus struggled for terms in which to describe the "power and the nature" of this prior reality, which is, as he puts it, "of all becoming the bearer and the care-giver" (49a).[1] It is a matrix always in motion, never a static eternal "this" or "that" but with endless generative capacity. Not inert or without form, it has "powers" of its own to give birth to the endless variety of elements and combinations of elements that make up different physical substances (52d–e).

> Its motion, like an instrument which causes shaking, was separating out farthest from one another the dissimilar, and pushing most closely together the similar; from which also these Kinds occupied different places even before the Universe was organized and generated out of them. (53a)

Indeed, Timaeus goes on, although the universe was not yet in any state of "reason or measure" and things were disposed as they are likely to be without God, fire, water, earth, and air already possessed traces of their own nature before any Creator marks them out into shapes by means of forms and numbers (53b–c).

In Timaeus's account of this prior nature, there are no rigid eternal structures, no fire that is never anything but fire or water that is always liquid, but neither is its "shaking" motion chaotic or random. Out of its natural mixing and mingling, physical substances form and reform, not as fixed beings but transmuting, changing, dissolving, defusing one into another to make what is more or less solid, more or less fiery, airy, or liquid. Water congeals and becomes solid, earth melts and erodes, one element becomes another in processes in which there are no hard-edged unchangeable beings, only a becoming, converging, changing reality (49b–e). This originating "matrix" has a "necessity" of its own apart from the logical necessity of mathematics, a necessity that reason must take into account. In effect, said Timaeus, the cosmos in its origin was generated as a "collaboration" between necessity and reason, and in that collaboration, reason has no absolute power. It must work to "persuade" nature to arrange things according to its dictates; it must yield as well as command.

Acknowledging that he is only able to give a "probable" account of the relationship between these aspects of reality, Timaeus tries several metaphors: The Creator is like a perfume maker who chooses for his base an odorless substance so that the fragrances he adds will not be altered. Nature is a "maternal" matrix that takes on "paternal" form, resulting in blurred imprinted offspring as natural kinds (50d–e). And finally, the third kind is a "χώρα" or location, eternal and admitting of no destruction, a place or εδρα for all things that come to be apprehended only by a kind of quasi-reason (52a–b). Barely capable, if capable at all, of being an object of rational philosophical knowledge, but understood as having to be there given that anything that exists must be "somewhere" and occupy some space, of supposedly little reality compared to the two entities discernable by the philosophical mind—Forms and the blurred outline of Forms in physical objects—nature refuses to dissolve into nothingness. As Timaeus put it, "We dimly dream and affirm it as necessary that all things that exist should exist somewhere and occupy some place, because if a thing is not in heaven or on earth, it is nothing" (52b). Timaeus's χώρα is not empty Cartesian extended space or a characterless "container"; it is the matrix of reality, the socially bounded natural places and habitats in which real events, persons, and things are situated.

The obvious objection was raised by Timaeus himself.

> Who is to say that this moving generative nature is not all that there is? Does fire-itself, or any of the other things we call always being themselves, exist, or is it only the things that we see, or otherwise perceive with the bodily senses that truly exist, and nothing else exists anywhere or in any way, and it is no more than an arbitrary pronouncement of our own that there always exists a distinguishable form of everything, no more than a string of words? (51b–c)

Why not say that this natural matrix is reality? Why posit a higher, a more truly real Being of eternal form? He, for one, would vote for Form, said Timaeus, but some sort of argument has to be made and he will make a try at it: Either there is a difference between reason and true belief and there must be Forms as imperceptible objects of rational thought; or there is no difference between reason and true belief, and the only reality is what is perceivable by the senses. But true opinion and philosophical reason *are* different. True opinion is formed and altered by "persuasion" and possible for everyone. Reasoning, on the other hand, comes from instruction, is rational, not alterable by persuasion, and possible only for gods and a few men. It follows that there must be Forms, because if there were no forms there would be only true opinion. If reasoning is to exist as the province of an elite class of rational thinkers, so must a realm of Being exist that is fixed, eternal, immune to persuasion, and beyond the give and take of ordinary deliberation, inquiry, and debate. Whether Timaeus meant the argument—if we philosophers are to have anything to philosophize about eternal forms must exist—completely seriously is hard to say.

Four hundred years later, reassuring his friend and fellow priestess Clea that her family's initiation into the mysteries of Isis did not conflict with the Platonic ideals of Delphi, Plutarch appealed to Timaeus's revised account of primal reality. Look, he told Clea, for complementary symbolic meanings and deep truths of natural and human life in religious rites and rituals. Look for common themes found in many religious traditions. If the Egyptian Osiris is comparable to the Reason of Apollo, the Egyptian Isis is the female nature, matter, or material that preexists and is given form by the divine Logos in Timaeus's Platonic account of creation (*Isis and Osiris* 371c7–15). Like Timaeus's matrix, Isis is not inert, not without force of her own. She actively seeks out Osiris, and when she finds him, she must be persuaded to lie with him. Like Diotima's Penia, Isis is in no way raped, impregnated, or mastered against her will. She is needy, actively seeking, longing for resourceful intelligence. Nor is she active only in initiating a primal coupling with reason that results in cosmic order. After the world takes form, some force or power must act to maintain and repair that order from the destructiveness and chaos of the Egyptian Typhon, the personified forces of disintegration and violent conflict. When Typhon overcomes Osiris, when chaos overcomes order, Isis as nurse and mother must be there to preserve and repair the harmony that has been damaged or lost. She must intervene as "a moving nature, both living and intelligent" until order is restored (375c).

* * *

In Christian centuries to come, it was Timaeus's first unrevised account of creation that would prevail, weighing down monotheist thought and creating endless contradictions and anomalies, many of which had been recognized by Timaeus himself. The Creator must be "good," which means he must also be "intelligent and "rational." However, as noted by Timaeus, copies of eternal form that make up the created visible world are inevitably imperfect, as is any visible copy of a schema. Does that mean the Creator is less than powerful? If "being good" he is not "jealous," and not being jealous he creates the world in his image, why are the world and the creatures in it not immortal like himself? Why must they die and suffer? Why create disease and pain? And why are the humans he designed guilty of such unspeakable acts of violence? For Socrates and friends gathered for a second day of philosophical entertainment, the questions were subject for debate. In the context of doctrinal institutionalized Christian belief, they had to be answered decisively.

One recourse would be ineffability. Avicenna combined Platonism with the logic and metaphysics of Aristotle to theorize the Creator God of Western monotheism as a unique necessarily existing Being with no knowable differentiating attributes. The Jewish philosopher Maimonides provided a "Guide for the Perplexed," for those struggling to understand the God of the Hebrew Bible in rational terms. About God, said Maimonides, nothing can be said. As the prime mover, and the final and efficient cause of all that happens, he can be known only indirectly through study of the operations of nature. Anthropomorphisms like "I am a jealous god" or "God is angry" were devices used by prophets and intelligent leaders and rulers to help less rational masses control their impulses. For Thomas Aquinas, God was not one kind of Being, along with other beings whose existence is contingent and a matter of fact, but a Being that combines existence with essence as the ground and first principle and cause of all Being.

In science and mathematics, the eternal Being of any one rational structure of the universe soon faded from view. Non-Euclidean geometries modeled new multidimensional spaces. The nested hierarchical oppositions of dialectical logic splintered into quantification theory and semiotic iterability. By the nineteenth century, anything like the fixed geometric mathematical world order dreamed by Platonists was gone, replaced by a generative reality more like the dynamic processes of Timaeus's matrix, Plutarch's Isis, or Empedocles's philitis. Quantum mechanics revealed a seething creative mix of emergent microevents rather than an ordered universe ruled by natural law. Biologists studied human beings as changing, evolving organisms rather than souls clothed in designer physical bodies. Species were no longer fixed as "natural kinds," but were understood as dissolving and developing in complex interactive processes of inheritance, environmental forces, and goal-driven choices.

How to maintain a vision of God as the all-powerful divine Creator of a good and rational universe was the challenge of generations of European and American theologians. Otto von Harnach, Karl Barth, Martin Buber, Reinhold and Richard Niebuhr, and Paul Tillch lived through the demise of Newtonian physics and the catastrophic violence of two World Wars. Could any version of faith in such a God be maintained without recourse to credulous biblical literalism? Was any kind of supernatural Being anymore believable? The key, said the Jewish theologian Martin Buber, was to strip God of "objective" qualities, to understand that God is not any kind of object Being, but rather an eternal Thou that revealed himself to biblical prophets and can be glimpsed in nonobjective personal encounters of one person with another. For the most part, argued Buber, we experience others—kings, princes, lords, lovers, legislators, designers—as objects to be categorized, manipulated, or managed, but at rare transcendental moments, we meet another person not as an object but as a "you" in relation to "me." Always there is the danger of reverting back to experiencing persons as objects, but, argued Buber, in each rare genuine I/Thou moment, a little bit of God is experienced as the "absolute Thou," the "thou" that can never become an "it," that cannot be inferred from any given revelation or cosmic design or understood as any idea, Form, or thinking substance. In "consummate events" of person-to-person relation, we glimpse an eternal disembodied Being that can never be objectified.

To think of God as a "person" with whom one could have a "personal relation" was already too objective and anthropomorphic, countered Paul Tillich in *Systematic Theology*. God cannot be someone you happen to meet or someone with whom you could have a conversation as if God were a human person. If God cannot be an object in the world like other objects, he cannot be a person either. Nor can any Stoic Logos account for the "fallenness" and sense of guilt that is at the heart of Christian experience. Instead, "theism" must be transcended in a modern-day Protestantism that does not require belief in the Being of a personal God. God, said Tillich, is not any kind of being, person, or object, but rather the "ground of Being" and, as such, the source of an existential "courage to be" threatened in a modern age of anxiety and loss of meaning. The "courage to be," said Tillich, is a creative "eros" expressed in the work of artists and lawgivers and participation with others as members of a community (*Courage to Be* 47). But Tillich's courage-giving community was not Diotima's coming together in cooperative union of lovers, domestic households, and democratic assemblies. It was the Christian church with its ancient symbols, traditions, sacraments, education, and authority. In the end, admitted Tillich, even diligent religious practice could not deliver the "salvation" promised in Christianity. In the depth of modern despair, rationally sure that Providence is mere superstition and immortality is a figment of one's imagination,

Christians had to take a "leap of faith" into the unknown and believe in a "God above God," a God above any kind of rational belief.

By the late twentieth century, poststructuralist postmodernist critiques of the "logocentrism" inherent in the Western philosophical tradition from Plato to Heidegger had called into question the status of any religious belief. In the wake of Jacques Derrida's "deconstruction" of any lingering postplatonic logical discursive order and postmodern attacks on the validity of any eternal ahistorical truth, a nihilistic abyss opened in the form of a swarm of diverse styles, ways of thought, and eclectic movements in which true or right was no more than the "might" of whatever institution or person happened to be in power at any given moment. In the turmoil of 1960 and 1970's politics, the critics of Platonic Being struggled to find grounding for any committed ethics or coherent political praxis. It could seem the only recourse was resistance and transgression, a continual anarchic shattering of conceptual idols, a breaking up of any discourse that pretended to authority, the undermining of any social or political order with the power to enforce its decrees.

With the great edifice of what philosophers since Aristotle had called "metaphysics" reduced to rubble, some remembered Timaeus's forgotten χώρα, reconceptualized as a vast echoing emptiness. Take away God as Platonic Being, as the oneness of logical form and the supposed outline of that form in natural kinds, take away any steadying symbolic, linguistic, or discursive order, what is left was Timaeus's illusive mysterious "third." Heidegger gave the signal, mentioning "*khora*" in his *Introduction to Metaphysics*, along with the suggestion that this might be something prior to the modern alienated concept of space as extension (70). Julia Kristeva followed, identifying it as a paralinguistic expressiveness that underlies symbolic and semantic structure (*Revolution in Poetic Language* 26). Derrida referred to it as a mysterious pretextual "differance," a kind of sieve or primal spacing or iteration that exists before signification and "between" sensation and intelligibility.[2] For post-Derridean philosopher John Caputo, Timaeus's "*khora*" was simply nothing:

> *Khora* has no meaning or essence, no identity to fall back upon. She/it receives all without becoming anything, which is why she/it can become the subject of neither a philosopheme nor mytheme. In short the *khora* is *tout autre*, very. (*Prayers and Tears* 36)

Inevitably, ghosts from the monotheist past returned to haunt that supposed empty "otherness." Late in life and facing serious illnesses,

Derrida heard fragments of Old Testament prophecy from his Jewish childhood, voices that he confessed he heard even as he proclaimed himself an atheist. He meditated on the story of Abraham's sacrifice of Isaac.[3] Like Moses, he experienced a sensation of being "called" or summoned by something beyond or behind ordinary experience. He imagined an unnamable presence uncovered or freed by the endless textual work of deconstruction that had not in the end been silencable. He had premonitions of an unexpected, impossible, miraculous unknown future happening with no foothold in logocentric reason—a child born to an octogenarian, a virgin birth, an earthly kingdom of peace and justice. Disconnected phrases from early religious instruction surfaced—a transcendent "gift" of "love," a "promise" of messianic deliverance from bondage—cited, quoted, removed from historical or scriptural context.

Jean-Luc Marion's Catholic version of a postmetaphysical God in *God without Being* reverted more specifically to the Lover God of the Hebrew Bible. Marion's God "reveals himself" with inhuman "excess"; God "offers himself" to be contemplated; he "gives himself" to be prayed to. God gives his love in a mystical marriage, instantiated in the body-to-body intimacy of the Eucharist and performed in the "vow" of profession of faith that "Jesus is Lord." In *The Erotic Phenomenon*, Marion described a mystical passion of intimacy with a jealous God. He rejected the dissection of love into irrational concupiscence on the one hand and purified gratuitous agape on the other. To understand divine love as "the thing itself" without preconceptions, he chronicled his own first-person experience of stages and figures in passionate relationship, speaking, as he put it, in the name of "others" as well as in the name of God (9). It is from the biblical God, he said, that we must learn to love, the God who cried out "Does anyone love me?," the God who loved first without knowing the people that he loves or whether they merit love, the God of Isaiah and Jeremiah who makes demands, is jealous, becomes flesh, and conceives a child Jesus (222).

Marion's prophetic phenomenology of stages of human and divine love bears little resemblance to the work of love as explained by Diotima. If for Diotima love begins as need, it is not the need of an isolated ego seeking validation of his own worth and asking "Does anyone out there love me?" but rather an enterprising sense of lack that looks not to self-validation but to deliverance of the heavy burden of mortal capability. Eros for Diotima is not, as Marion describes it, a cry for recognition or the arbitrary choice of someone to love before you know who or what she is, but rather response to the fact of blooming or budding of beauty in another person that presages productive interaction. The aim of eros is not a mutually validating "gaze" or a carnal coming together but good things in the world that come from physical and psychic interaction: sensual pleasure, children, homes, conversation, common cause. Marion's God is the jealous, demanding, relenting divine husband of prophetic scripture, who gives to his undeserving and unfaithful people the gift of a son.

In these modern and postmodern theologies, there is none of the creative stirring of Timaeus's matrix, but rather broken fragments from the past, memories of a God imagined to appear miraculously, reveal himself, promise, and mingle intimately with a struggling marginalized people. Why is Tillich's supposedly monotheist ground of being "God above God" if it is not Yahweh proclaiming his superiority to the other gods of Canaan? What is Buber's divine "Thou" if not the ghostly voice of Yahweh, speaking through prophets condemning pagan worship and promising victory in battle. Why must a believer cry out and "wait" for God to decide to "reveal himself" or "disclose himself" as if God is a potentate deciding when and where he will make an "appearance" and whom he will love? The dim shadow of such a God is all that is left of the "naturalized" Christian God proposed by contemporary philosopher of religion, Mark Johnston, in *Saving God*. If "religion" is to survive in an age of science, he argues, it cannot be "held hostage" to irrational belief in resurrection and personal immortality. If God is to be "saved" as the "Highest One" of Scripture—the one that could not intend evil or have contempt for the truth—God must be reconceived in some noncredulous way.

Given that in many passages of the Hebrew Bible, Yahweh appears as a murderous jealous patriarch, why do we need such a God or indeed any god at all? For Johnston the answer is an essential human self-centeredness that makes goodness all but impossible to conceive without divine sanction. As "fallen creatures," "compromised by self-will," we can only be saved by something outside us. Because we perceive the world egocentrically, neither law nor the "idea" of goodness can make us moral. Obedience to law can be no more than a way to avoid moral choice; goodness can be no more than behavior that allows us to avoid the condemnation of others. If true goodness is to be possible, there must be something else, something—a God—that is not nature as in Spinozan pantheism but something that reveals or discloses or manifests itself in nature. Johnston searches for words to describe this indeterminate "something" or "someone" that is not the soul of the universe but rather something like its "form," or "intelligibility," or the "totality of laws of nature," or "the sensory presentation of things." Again echoes sound from the past: the Stoic logos, Thomas's created aesthetic form, "revelations" of Christian saints, and, as an add-on, a Buddhist selflessness that might make the Christian mandate of "love your neighbor as yourself" possible. Eros is summarily dismissed as of no use. Episodic, limited, of no moral value, declares Johnson, it is a dead end. Love-stricken couples like Tristan and Isolde suffering from "hysterical identity confusion" can hardly be taken as examples of virtue (*Surviving Death* 347–348).

Even more decisively untethered from biblical faith is the "religious atheism" professed by Ronald Dworkin in his last work *Religion without God*. Stripping away the untenable "science" of biblical religion—the short age of the universe, virgin birth, providential history—Dworkin retained a core of belief rooted in person's individual sense of moral responsibility. Simply to think, argued Dworkin, that how I live my life is important and that there is deeper meaning and wonder in the universe is "religious." At the end of a life's work in practical ethics meticulously analyzing and resolving conflicts between self and others in contemporary moral issues, Dworkin professed a faith that was in no way theistic. It matters whether a person supports same-sex marriage or capital punishment. The fate of the universe and the future of humanity is important. You do not have to believe in God, said Dworkin, to have a sense of value, mystery, and purpose in life, or think that nature is not just a matter of particles thrown together but something of wonder and beauty. Beauty is there in the world, whether or not God created it; moral goodness and virtue is valuable regardless of religious commandment.

Although Dworkin's professed "atheist" religious values come close to Diotima's "work of love," family harmony, civic responsibility, and appreciation of natural beauty are detached from the daemonic spirituality that in Diotima's teaching supports, inspires, and maintains these human goods. Once you reject faulty biblical science, said Dworkin, there is no reason to engage in any religious rituals, worship, or ceremony. Rigorous rational argument can be relied on to prove and validate moral judgments supported by consistent systems of moral principles. The skeleton of Platonic Being hovers in the background, stripped of Old Testament patriarchy and messianic hope, but still dependent on the highroad of Parmenides's goddess of Truth.

The young John Rawls, writing his senior thesis at Princeton as an avowed Christian, was not ready to do away with the prophetic force of biblical revelation. Neoprotestants like Barth and Tillich were right, said Rawls. A rationalized Platonic Logos supported by logical arguments cannot replace the moral force of the Judaeo-Christian message. But nor can a secular "eudaemonist" ethics that takes human happiness as the highest good, or an "ethical naturalism" that reduces morality to individual desire and appetite. The answer, said Rawls, is Christian *agape*, divine love that binds together not only a church or a nation, but the Christian community, flowing out to neighbors, citizens, and people of other nations. The limits of his youthful communitarian theism were clear to Rawls returning from military service in World War II. His experiences in the war, he said, along with extensive reading in the history of Christianity had forced him to reject his early optimistic Christian faith. However rationalized or refined, original sin, heaven, hell, salvation by grace, and predestination made God out to be a monster in which few people can or should believe. Christianity, far from having a moral or socially beneficial effect, should be completely separated from politics.[4] Rawls concluded his postwar essay "On My Religion" with what he called a "naïve question": "Is it possible to imagine a noncredulous religion that does not challenge scientific reason, that does not lead to sectarian dispute or holy war, but that supports the best in human endeavors?"

* * *

There is no sign in Rawls's later work mapping out a secular theory of justice that he ever found an answer to that question, but Diotima's lessons in love provide a possible beginning. Divinity, as she describes it, is not a person or object of any number, sex, or kind, but a current that circulates throughout the natural world. Without the expectation that a god might "reveal himself" to chosen humans, without the insistence that God is one or many, powerful or loving, or any other attribute or set of attributes, divinity is experienced as the real attractive and visible possibility of transcendent beauty and happiness. There is no transport to heaven behind the chariots of gods as in Plato's *Phaedrus*, no direct communication with Yahweh as in the apocalyptic visions of the biblical Daniel. What is plural, and knowable as plural, is not gods but "spiritual" or daemonic avenues of communication between what is and what is beyond and transcendent of what is. As described by Diotima, these include, along with eros, priestly rites, prophetic and mantic revelation, sacred music, celebratory processions, and festivals. In many of these activities, images are involved, not as idols or likeness of specific divine entities but as points of symbolic meaning and focus. The beauty in bodily form that arouses the sacred first transcendent pangs of eros can be represented in the statue of a young Apollo or Adonis. A symbolic tree or pillar may stand for Asherah as the creative fertility that is the immortal spark inherent in nature. Inanna is celebrated as the protective spirit that guards the common storehouse of a people, ensuring the survival of a community in times of scarcity. A mound deep in a crevice may be sacred to Eileithyia or childbirth as the dark hidden nurturing of new life. Sacred images do not function as pictures of divinity like representations of Jesus on the Cross in traditional Catholic belief. They do not demand fidelity. There is no sacrilege in adding the name of Yahweh to a hymn to Asherah, or placing an Asherah column in Solomon's temple. All represent pathways to the divine, all may be media for divine inspiration.

Without the objectification of divinity, there is no need to distinguish God from his fallen demonic agents or decide which God is most high. There is no need for hierarchies of being. There is no contradiction in visiting a mountain shrine to Aphrodite, praying to Mary in a Christian chapel, and making an offering in a Buddhist temple. Hera, Apollo, Aphrodite, Moira, Eileithyia, Dionysus are not separate entities but visible images and aspects of multifaceted religious sensibility as described by William James in *The Varieties of Religious Experience*: Behind all the many forms of religious practice, there is a core of religious sensibility: a feeling of uneasiness and lack, the sense that there is something beyond one's own mortal life, the belief that the visible world is part of a greater spiritual reality, the conviction that harmonious relations with that reality are the "true end" of human life. In this core sense, prayer, ritual, inspired prophecy, and other religious practices bring joy, heroism, a zest for life, and a sense of safety and purpose.[5]

Monotheisms inevitably lead to the violent conflicts that Rawls found so disturbing in the history of Christianity and so disruptive of community and regional harmony. Objects and persons, natural or supernatural, are by definition existent or nonexistent, making arguments for the existence of God vulnerable to skepticism on the one hand and irrational blind "faith" on the other. Whether God is imagined as a ruling Lord of Hosts, a jealous husband, or an objectified reason or Logos, antimonies in theological thinking inevitably undermine belief. If God is all-powerful, how is freedom of the will possible? If God is provident and arranges the universe for the best, how is it possible to understand the horror of war and violence committed in the name of religion? Once belief in God becomes dogma, how can religious authorities be prevented from persecuting unbelievers or rejecting useful theories in science because they conflict with theological belief?

No contemporary theologian addresses these questions with more urgency than Catholic theologian Elisabeth Johnson in *She Who Is* and *Quest for the Living God.* Johnson asks: Is there any way to talk about divinity that is not now suspect? To talk of God as a masculine warlord, legislator, or father has consequences for morality and social organization no longer tolerable. But to simply reverse genders, add female attributes, and talk about She or Mother instead of He and Father does little "to create conditions for the formation of community characterized by relationships of mutuality and reciprocity, or of love and justice" (*She Who Is* 6). Simply changing the gender or titles of God may not "serve the emancipatory praxis of women and men" or save the natural environment from ecological disaster (8). Even if one could find for God some neutral nongendered denomination like "ground of being," "holy mystery," or "eternal thou," there may already be something presumptuous in any attempt to identify one true and accurate image of divinity. Many of Johnson's examples are drawn from the pre-Christian past. Ishtar, she notes, did not have exclusively male or female powers. Like Aphrodite, she was depicted as warrior protector of the community, as well as nurturer in birth, fertility, and care of the young. Early multidimensional divinities included all the many beauties—male, female, animal, and other—of being in the world (69).

If descriptive words for divinity cannot be taken literally, they can be understood as pointing to spirit as transcendent and nonobjectifiable. In *Quest for the Living God*, Johnson describes that spirit in terms recalling Timaeus's matrix and Empedocles's philitis updated with insights from contemporary science.

> [Spirit is] creative power who dwells at the heart of the world, sustaining every moment of its evolution...the physical world itself is the matrix of God's gracious indwelling...Bringing the Spirit back into the picture this way leads ecological theology to envision God not at or beyond the apex of the pyramid of being as in modern theism, but within and around the emerging, struggling, loving, dying, and renewing circle of life and the whole universe itself...The universe, in turn, is self-organizing and self-transcending, energized from the spiraling galaxies to the double helix of the DNA molecule by the dance of divine vivifying power. (188, 189, 91)

On this view, life and mind require no supernatural intervention. Like Plutarch's Isis, Johnson's Spirit works within nature "bringing forth and nurturing life, holding things together, renewing what the ravages of time and sin break down." Like Lucretius's Aphrodite, Johnson's Spirit is expressed in the renewal of spring, the network of life, the recreation of human culture, and erotic love.

> Fertility is intimately related to her recreative power, as is the attractiveness of sex. It is she who is ultimately playful, fascinating, pure, and wise, luring human beings into the depths of love. (*She Who Is* 135)

Johnson's "living god" is experienced in specifically Christian communities and in sacraments like the Eucharist and the confession of faith, but even more vividly in practical experience in human communities actively struggling with the ugliness of violence, injustice, poverty, and environmental damage. Her examples include young German theologians after World War II coming to terms with the horror of Nazism, liberation theologies in South America, the contemporary women's movement in the United States and Europe, the fight against apartheid in South Africa. In each case, she describes a progression similar to Diotima's work of love. First comes experience in a community recoiling from ugliness and responding to promising "buds" of beauty in victims, coworkers, and allies. Then comes thought and reflection, both academic and popular, which in turn lead to continuing committed action in the world aimed at the creation of just institutions and laws. In these cases, said Johnson, one does not talk directly about God; it is as if divinity is glimpsed "out of the corner of one's eye" (*Quest* 2). You do not begin with theology or with the idea of God as Logos, or Goodness or Being, and ecstatically contemplate and worship that idea; you begin with engagement in human situations out of which come conceptions of divinity that change with ongoing experience in a particular material situation.

Spirituality in this sense is not limited to one religious community or belief. In this, Johnson draws on the thought of the Catholic theologian Karl Rahner (*Quest* 28–48). Rahner, she says, began not with one primal divine design or providential plan for history, but rather with the human subject who as mortal feels an urge for self-transcendence that in turn presupposes a transcendent goal or "horizon." Anyone, Christian or not, who experiences and acts on the pull of aspiration experiences this divinity. As Johnson puts it, "the diversity of the world offers fragments of beauty, goodness, and truth, both social and cosmic, facets of reality that point us in different ways to the *one* ineffable source and goal of all" (*She* 118). The crucifixion of Jesus is a unique historical event, but, in the spirit of Plutarch's syncretism, in that event Johnson found universal meaning not that different from revelations associated with the mysteries of Demeter or rites in honor of Dionysus. Death and suffering are inevitably part of mortal life, but out of death comes regeneration and rebirth, the realization of which can awaken resistance and revive hope. In the story of a God who allows his son to suffer for others appears the image of the love of mothers like Alcestis who, as Diotima noted, are willing to die to save husband or children. The incarnation of Jesus provides the image of a father who sends his son to be the savior of others, so that "the universe is seeded with promise" (*Quest* 191).[6] Biblical stories of "promises" to a struggling marginal people—that with effort and resource they will find a safe homeland—provide motivation and inspiration that spurs on resistance to oppression and energizes new initiatives and undertakings.

On this view the generative spirit that sits at the center of the world is not "loathsome" as taught by Parmenides's goddess of truth, but divinity itself, working in the world in ways that do not require supernatural interference with natural order. "Creation" and "providence" are not ways God imposes his "will" on the world, but ways the world and its creatures are empowered to create themselves. What results, says Johnson, is neither a rigid determined order of Being nor random orderless chaos, but the observable fact of "emergent, adaptive, self-organizing systems."

* * *

Along with belief in heaven, hell, and creation, prominent among the religious ideas regularly cited as pernicious and damaging to rational scientific thought is belief in supernatural agency. A common narrative in standard histories of Western culture is the supposed battle between irrational religious "faith" and scientific truth. As the story goes, religion—anchored in biblical faith, rationalized by Aristotelian physics and Platonic metaphysics, institutionalized by the Catholic Church—held back the pursuit of knowledge through the medieval period. When advances were made in scientific understanding in the early modern period, the authority of the church stood in the way of the "enlightened" struggle of newly revived scientific reason against superstition. On this view, religion is an "illusion" or "delusion" that continues to bedevil and impede true "scientific" knowledge. It might have once seemed that priests could change a substance miraculously into another substance, or that organic form requires a supernatural designer, or that high in the sky exists another heavenly realm. But, argue religion's scientific critics, this is pure fantasy in an era in which atomic theory, evolutionary biology, and astrophysics prove conclusively that all change can be explained by the mechanical movements of elemental particles, that humans evolved from nonhuman primates, and that the earth is a speck of dust in an expanding universe doomed to grow increasingly cold and lifeless. Only those who refuse to be rational could anymore continue to believe in heaven, creation, or God. In response, defenders of religion proclaim ever more vehemently the truth of biblical teachings and the danger to morality and society in any science that rules out acts of God.

In these polemics, "religion" is understood as the one religious tradition of Judaeo-Christian-Islamic monotheism. As with Diotima's "poesis," the use of the word "religion" narrows to just one application, and the word's more general meaning as spiritual values and beliefs that bind a community together is lost. Early tribal cults, nontheist religions like Taoism or Buddhism, the many varieties of animism, all are dismissed as no more than "primitive" beginnings of religion, utterly superceded by Judaism, Christianity, and Islam.[7] The great variety and sophistication of Mediterranean cults and beliefs outlawed under Constantius II in the name of imperial Christianity are bunched together under the dismissive epithet "paganism," coined by Constantine's army for the beliefs of naïve "country folk" incapable of appreciating the power of the Christian standard in battle. On the other side of the debate, science is understood equally narrowly as atomistic physics, neo-Darwinian evolution, and the big bang theory of the origin of the universe in an exploding mix of elementary particles. Given these two alternatives, unless one is willing to live with irreconcilable contradiction, the answer to the question whether it is possible to conceive of divinity in an age of science must be no. There is no way to conceptualize transcendence in a world in which everything that happens is the result of the mechanical repulsion and combining of elementary particles according to invariant laws of physics, and in which the behavior of living beings can be fully explained by chemical and neural processes. Even less anthropomorphic forms of Christian divinity like Tillich's ground of Being or Buber's transcendent "Thou" must be rejected in a world where there are only mechanical processes in which patterns emerge, congeal, disintegrate in a proliferation of chemical or organic structures. In such a world there is no room for existential grounding or mystical recognition of "the Other." Spiritual experience itself can be explained and dismissed as the effect of a mutant "god gene" that orchestrates a firing of neural pathways in the brain that happen to have once had survival value for human organisms.[8]

As is the case with any fight, the battle between religion and science takes place on common ground. Both theists and the scientific critics of theism share a core belief: beyond the perceivable world is another eternal order, whether in the mind of a powerful Creator god or in the form of invariant physical law. Physicist Lee Smolin points to the lingering Platonic and religious yearning for an eternal "law of everything" that gets in the way of understanding in contemporary cosmology:

> What I have been arguing is that the presumption that that which is timeless is somehow greater, more real, or more true than that which is time-bound is essentially a religious idea that, sooner or later, comes into conflict with our desire to have a rational comprehension of the world in which we find ourselves. It points always to an imagined world, and with it an imagined intelligence that exists somehow outside of our world, free of the bounds of time that life and death impose on us. (*Life of the Universe* 210)

But new paradigms have replaced this older view of a universe ruled by universal law. As Smolin puts it: "Thus the metaphor of the universe we are trying now to imagine, which I would like to set against the picture of the universe as a clock, is an image of the universe as a city, as an endless negotiation, an endless construction of the new out of the old" (*The Life of the Cosmos* 299).

A similar argument in environmental science is given by Daniel Betkin in *Discordant Harmonies.* Mythic and theist ideas, Betkin argues, distort scientific thinking in ecology into the twentieth century. Ideas of a static God-ordained order of nature and of man's special status in relation to nature carry over into and subvert the methods and mathematics of ecological analysis, whereas in effective environmental regulation and practice, what must be maintained is nature's own "reproductive capacities."[9] Similarly in biology, Evelyn Fox Keller's work with aggregate versus individual behavior in slime molds was rejected on the faulty premise that if individuals worked together in concert, it had to be because directing pacemaker cells were giving orders. Although at first the idea of the collective action and emergent formations of interacting cells seemed incomprehensible and even mysterious, her work has eventually led to studies of self-organizing relational behavior in many different areas of biology.[10]

Hard as it might be to reconcile any version of theocratic design with mechanistic atomism, these advances in evolutionary theory, networking technology, biology, physics, and other sciences are not only consonant with the daemonic spirituality described by Diotima but inspired by it. Nothing in Diotima's account of divinity requires a designer or creator external to nature. Transcendence and creativity are aspects of natural life and especially human life. Change in climate, habitats, organisms, social organizations, art, or knowledge are not driven solely by determinate microevents. Living creatures feel lack and deficiency, shrink back from ugliness, perceive beauty, are drawn to it, and enter into productive relations that result in achievements that survive their own mortal existences. "Evolution" of new forms of life is not a "mindless purposeless algorithmic process" in which a few of many random combinations of genes survive in particular environments but rather in part the result of purposeful cooperative activity. Since the discovery of genes and the first neo-Darwinian syntheses of Darwinism and genetic theory, theories of evolution have become more complex. In nonhuman animals, sexual desire is no longer understood as hard-wired automatic seasonal response to organisms of the opposite sex. Female animals are courted by males, treated to complex displays of beautiful plumage, impressive horns, and feats of strength and fitness. In most cases, females choose to accept or reject sexual overtures. If, in animals, genetic inheritance is not a random throw of genetic dice, but a result of perception of beauty and aroused desire, choice is even more evident in human sexuality. Eros, in the form of selectivity and pair bonding, is an important evolutionary basis for human society according to recent research in primatology, accounting for the crucial advance from the crony-male dominance and orgiastic promiscuity of chimpanzees to social forms of reproductive life and the achievements of civilization.[11]

Nor can genetic inheritance be any more understood as rigidly determining mental or physical characteristics. Humans grow, change, and develop in communicative relations with other beings, with the world around them, and with themselves internally by way of the "forgetting and replacing of ideas" identified by Diotima as "studying and learning." In the process of living, individuals, the others with whom they interact, and environments change, change driven in part by inherent desire, purpose, and design. As one contemporary biologist puts it: "We all know the strength of our yearnings, those inchoate desires that nag at us and goad us to fulfill them, to shape the world into something imagined even if we cannot articulate precisely what the desire is" (Turner, *The Tinkerer's Accomplice* 182).[12] On this view genetic inheritance is a sketch, worked and reworked in body and brain, not randomly but driven by lack, need, and clever design. Not only is there a "fundamental intentionality" in the behavior of living things, the designs of living things outlive them and survive for generations, permanently changing the course of nature. If this is true of nonhuman organisms like bees, ants, termites, or beavers, it is even more true of the progressive movements cited by Johnson and the cultural works of human inventors, artists, lawgivers, and producers of knowledge cited by Diotima. Life, understood from this scientific perspective, is never individual, never a matter of lone individuals competing for control of some environmental niche. It is interactive, relational, and dynamic with living things functioning in harmonious and disharmonious relations with other interconnected living things as well as with the inanimate environment. Nature is neither a giant organism, as in one strain of modern thought, nor a collection of aggressively competing individuals, but rather networks of affiliation and cooperation that generate new ways of living.

* * *

A perennial issue in debates between science and religion has been the status of morality with defenders of religion arguing that morality must come from God and be rewarded in heaven and defenders of scientific materialism claiming to show that the origin of morality—defined as altruism opposed to selfishness as in Christian ethics—can be fully accounted for by natural selection. Again, the dispute takes place on common ground. In both cases, humans are seen as essentially self-interested, whether willful and "fallen" as in the Christian version or in the evolutionist version, driven by biological mechanisms that increase their—or in more extreme versions their genes'—chances of survival. At some point, the scientific story goes, whether in rudimentary form in primate behavior or later in Neolithic or Paleolithic eras, chance mutations occurred that induced one or more individuals to cooperate with other individuals with the result that they survived to reproduce more successfully. On this view, the Christian principle "Love your neighbor as yourself" is a version of "love yourself," a purely naturalistic development of competitive behavior that evolved over time and requires no spiritual intervention.

Much in this evolutionary account, however, is now subject to dispute. The existence of a primal "state of nature" in which solitary hominids compete for survival is highly speculative if not Hobbesian fantasy. There is no evidence of any hominoid creatures or perhaps any living creatures living in other than social groups. Although many different forms of erotic attachment exist in human societies—from pair bonding, to monogamy, polygamy, group marriage—affiliation of some sort seems an essential aspect of human life, and even of organic life. That individual self-interest is the driving force of organic life is more a matter of "faith" in the mechanics of mutation and natural selection than proven fact. Nor is it clear that human interests are limited to their own survival or the avoidance of pain. In the natural world, there are elements, patterns, forms, and symmetries that humans and many animals recognize as beautiful in other individuals, arrangements of plants, landscapes, fertile fields, harmonious social arrangements, and mathematical formulae. This along with clever intelligence that courts and wins lovers, friends, and allies inspires persons working together to create good things in land and resource management, law, and science. Morality in this sense requires no one Form of Beauty, no realization of a World Spirit, or apocalyptic end of history. This is not an order imposed on recalcitrant believers, not a divine template for a just society, not God's plan for the world. What is eternal is the fact of transcendence and the proliferation of new forms of beauty in creative interactive relations. Divinity is the fact of the constant purposeful reemergence of new forms of life out of generative relations fueled by desire.

Once one abandons the fantasy of travel to heaven and escape from dilemmas of natural existence, once one gives up faith in an all-powerful heavenly father or the search for an eternal "law of everything" governing the universe and fully grasps the implications of relativity and quantum mechanics, the universe looks much like the one Diotima and theologians like Elizabeth Johnson describe: a network of relations, subject to renegotiation and change, a composite of interacting perspectives and points of view, as well as emergent structures blessed with both regularity and novelty. Lee Smolin puts it in terms that echo Diotima's lessons in love:

> There may or may not be a God. Or gods. Yet there is something ennobling about our search for the divine. And also something humanizing, which is reflected in each of the paths people have discovered to take us to deeper levels of truth. Some seek transcendence in meditation or prayer, others seek it is service to their fellow human beings, still others, the ones lucky to have the talent, seek transcendence in the practice of an art. Another way of engaging life's deepest questions is science. (*The Trouble with Physics* vii)

Perhaps no one took that quest as much to heart as the mathematician and physicist Alfred North Whitehead writing in the aftermath of the dismantlement of the old Newtonian universe in the light of relativity theory and quantum physics. How is it possible to understand, not just the "mechanics" of quantum physics or the astronomical evidence for relativity theory, but what this knowledge tells us about ourselves and the reality in which we live? Struggling to fashion out of existing language terms with which to describe the new world of twentieth-century science, Whitehead described a metaphysics of events in process that in many ways is consonant with Diotima's teaching. Not only, he argued, does that metaphysics not rule out divinity, it is not possible to understand the world of contemporary quantum physics without it. At the same time, divinity can no longer be conceptualized as the Bible's "imperial ruler" or Aristotle's rational "unmoved mover," but must be understood as "the lure for feeling and the eternal urge of desire," "the conceptual realization of the absolute wealth of potentiality" that keeps the natural world in motion.[13]

CHAPTER 10

Social Virtue

Then and only then will it happen that, seeing beauty through what makes beauty visible and not grasping on to images of virtue, a person will give birth to true virtue.

—Symposium *212a3–6*

In Diotima's teaching, there is no one privileged way to be good. Many forms of religious practice—prayer, prophesy, incantation, healing, sacrifice—can inspire virtuous behavior, binding the two realms of human and divine together into one "cosmos." There can be beauty in injunctions passed on by religious authorities. There can be wisdom in the inspired utterances of prophets. At the same time, obeyed literally and without question, moral rules can become substitutes for virtue, false images, or εδϖλα of virtue. This is especially true when those directives are believed to come directly from a God imagined as the supreme and infallible ruler of the universe. When prophets or priests claim to speak for such a god, virtue can become obedience to law, and law instead of expressing a community's changing needs and evolving sense of justice can freeze into a network of prohibitions without utility or reason. Prophecy, prayers, sacred images, and ritual come under the control of predatory priesthoods, temple elites, and corrupt ecclesiastic institutions. Privilege masquerades as authority. Without eros and the ability to see and respond to what is ugly and beautiful in persons and in human institutions, religious ethics becomes empty posturing.

In the warrior cultures of archaic Greece, the lordly figure of Zeus, standing behind the keeping of oaths and the sanctity of host–guest relations, helped to mediate conflicts between feuding chieftains and keep Greece from slipping into anarchy. The danger of illusion is greater when one paramount God is thought to have communicated to chosen individuals' eternal and unchangeable commandments. Once the vision that inspired those ordinances is lost, what can result are false "images" of justice that turn neighbor against neighbor and nation against nation. The illusion of virtue as obedience to divine injunction is supported by the belief, rejected by Diotima, that direct human contact with divinity is possible. It is further enhanced by certainty that one's own image of God is the one and only God who revealed directly to privileged individuals' moral truths that stand regardless of "visibly" ugly or violent results. Holy war, the killing of captives, slavery, the burning of witches, persecution of homosexuals, extermination of heretics and infidels, all can be condoned on the authority of such a God. Feelings of revulsion at cruel, destructive, or ugly results are overruled; any tendency to bond with a prisoner or an enemy is "courageously" repressed.

In troubled early Christian communities where Jews mixed with Gentiles, Paul struggled with the rigidity and divisiveness of Hebrew religious law. Should non-Jewish Christians be expected to submit to circumcision? Should they be required to follow the strict prohibitions of Jewish dietary law? No, said Paul, righteousness comes through the grace of God and faith in Christ, not through obedience to law. Paul was not ready to give up the Lord of Hosts of the Hebrew Bible in his new role as the Father of Jesus, but he was willing to concede that some of the provisions of Hebrew law laid down by that authority might have no more than temporary authority. Some Old Testament teachings, he suggested, might have been meant only as παιδαγογός, or "guardian of youth," until the coming of Jesus, at which time Christians would came directly under the authority of God (*Galatians* 3:23).[1] Paul perceived the hypocrisy of much law-based virtue: follow the rules, comply with the letter of the law, have a clear conscience no matter the result. In Christian communities struggling with internal conflict and diversity, such a shallow conception of virtue was unlikely to keep the peace. Competition as to who would obey dietary laws most exactly, rumors about the sexual misconduct of rivals, hypocritical condemnation of trivial faults, resentment of those with spiritual power all created animosities that strict obedience to established law was unlikely to overcome.

In a famous passage on love in *I Corinthians*, Paul questioned the virtue of divine revelation and esoteric knowledge in the name of Christian agape.

> If I speak in the tongues of men and of angels, but have not love, I am a noisy gong or a clanging cymbal. If I have prophetic powers and

> understanding of all the mysteries and all knowledge or if I have the faith to move mountains, but have not love I am nothing. If I give away all I have and if I deliver my body to be burned but have not love, I gain nothing. (13:1–4)

The answer to conflict and dissension in the Christian community, said Paul, is not law but the love of God for man as evidenced in God's sacrifice of his son Jesus. If some of that love can reach the hearts of Christian believers, it can flow back to God and in turn beam some of its light back onto fellow members in the Christian community. In contrast, prophecy, the special gift of speaking in tongues, the choice to die as a martyr for Christ, the attainment of special esoteric knowledge set one Christian above another, creating rivalries and enmities. To be human, said Paul, is to "see in a glass dimly." Only after death could one hope to come "face-to-face" with God. In the meantime, he proposed humbler virtues of human interaction: patience, kindness, humility, civility, even-temperedness, bearing the burdens of others (*I Corinthians* 13:4–7).

But Christian agape, based on love of God rather than "seeing" and responding directly to the beauty in others, can itself become no more than a pose, allowing resentment and pride to fester beneath a show of obligatory affection. In another of Paul's letters, it is clear that civility and humility have been no match for men's resentment over the spiritual power of women in the Corinthian congregation. Women, Paul told the Corinthians, should keep silent in church. They should remain subordinate. If they need to ask anything, they must wait and ask their husbands at home (*I Corinthians* 11:2–16). The remedial edge to Christian agape is echoed by the contemporary Christian theologian C. S. Lewis in *The Problem of Pain*. How, asks Lewis, are we to understand the love of God for man that is to be the source and model for Christian agape? How is it possible to reconcile agape love with the fact of human suffering and pain? The answer, he says, is that agape, in God or in men, is not simple charity, kindness, or benevolence, but inflicts suffering along with joy. God's love is "persistent" in the same way an artist persists in improving a work of art to make it "better." It is "despotic" in the way a man loves a dog that he is training to be of service to him and no longer "wild." It is "provident" like a father's love for a son—not a "senile" father who tolerates the son's misdeeds but a patriarchal father who "uses his authority to make the son into the human being that the father rightly and in his superior wisdom, wants him to be" (*Problem of Pain* 37). Lewis's fourth and, he says, "noblest" analogy is between the love of God for Israel and Jesus for the Church and the love of a man for a woman.

> When we fall in love with a woman, do we cease to care whether she is clean or dirty, fair or foul? . . . Does any woman regard it as a sign of love in a man that he neither knows or cares how she is looking? . . . Love is more sensitive than hatred itself to every blemish in the beloved. (*Problem of Pain* 38–39)

Nothing, said Lewis, marks the difference between the pagan God and the Christian God so clearly as the contrast between a deity who moves the universe in the same way as a lover is moved by the beauty of a beloved and a God who plays the proper commanding manly role of "agent to patient, male to female, light to mirror, voice to echo" (41, 44). Lewis's divine agape love, whether between God and man or man and woman, is insistence on loyal service, assertion of paternal discipline, and the restriction of women to subordinate roles.

The infusion of philosophic rationality into biblical ethics by scholars like Philo might seem to offer more reliable moral results. When God is objectified as Logos and all humans are thought to have been created in the image of God with at least some of that reasoning faculty, logic and not "manliness" becomes the final arbiter of right and wrong. Reason might mediate conflict in communities of believers. Reason might claim precedence over conflicting local traditions and bring different cultures and peoples together in one true faith by rational argument. Such was the argument of Pope Benedict XVI in his controversial 2006 address to the Science Faculties at Regensburg. The rationality of Christianity, argued Benedict, allowed it to be spread by persuasion rather than conquest, in contrast with a faith like Islam spread "by the sword."[2]

The deification of forms of human reason, however, carries its own dangers of illusion. Virtue now depends on deductions from first principles taken as necessarily true based on concepts defined relationally in terms of hierarchical oppositions. A thought is true or false. An action is good or evil. The "betweenness" and "becoming" of daemonic spirituality disappears, and what is virtuous or vicious is based on the necessity of the law of noncontradiction. Concepts that must change as human cultures expand and develop are locked into fixed order, defined by the presence or absence of selected key features. Eros is animal, not human. God is one, and not many. A man is a true believer or a heretic. An action is good or evil. Given that, as Diotima argues, the reality we experience within ourselves as well as in others and in nature is in constant dynamic change, ideas and arguments by which morality is defined pull free from reality. Homosexuality is evil rather than good, unnatural rather than natural, a sinful willful choice rather than a naturally occurring human variation. Biblical teachings on creation are necessarily true regardless of empirical fact. Science and religion go to war, and the premises from which reason deduces moral consequences are based on prejudice masquerading as revelation.

"God gave the earth to men to use productively." This, argued democratic theorists like John Locke, is a "Law of Reason" written on the minds of men by God. But the conclusion that jurists in the New World deduced from Locke's premise—European agriculture was more "productive" than native American agriculture, therefore Europeans have a right to appropriate private property in the New World—was used to support predatory colonialist interests and defeat the rights of native peoples to ancestral lands. Looking back on the history of that process with clear vision, it is hard to "see" much virtue, and easy to see a good deal of ugliness. This is not to say that the idea of a divine Logos—a providential rational order of nature replicated in the reason of each individual person—cannot produce virtuous results. If every person has a spark of divine reason, every person is potentially a citizen of a kingdom of ends and every person should be treated with consideration and respect. With such arguments, the self-evidence of divine Reason has inspired democratic movements and insistence on "human rights." At the same time, distortions in interpretation and communication can result in different transmitted meanings. If "rationality" is the mark of a god-given human nature, its absence can be taken as a sign of moral deficiency. If those who do not speak Greek, or are of different cultures, or have darker skin color, or are female are not "rational" or fully rational as rationality is defined, the argument can be made that they must be governed by more rational masters.[3] Again, the danger to morality comes not from systems of ordered thinking, eminently useful and necessary in human life, but from identifying one or several of those systems with divinity. If one set of premises, one method of reasoning, one set of conclusions becomes absolute truth, what can result are false images of virtue—male supremacy, white supremacy, homophobia—rather than the "begetting" of true virtue.

Is it possible to combine the authority of law and reason so that one corrects the other? This was the hope of Kant, who was well aware of the dangers in revelation, the unreliability of speculative reason, and the embeddedness of human thought in time and space. There is no way to escape the human basis of knowledge or science in sensory experience, said Kant, but this does not mean that there can be no basis for moral obligation, but only human rules to be followed or not according to self-interest. The answer, said Kant, is in rational determination of moral law as universal principle. Regardless of priests or cults, human reason can recognize commands that are capable of being willed universally for self and for others and that supply the deontological certainty that revealed religion cannot. But again, results can be ugly. I will never tell a lie, never steal another's property, never kill myself no matter the consequences of telling the truth, the unfairness of the distribution of property, or the painfulness of fatal disease. Unnecessary harm is done, children go hungry, patients die lingering humiliating deaths. Kant's moral categorical imperative carries with it the form if not the name of divine monotheist commandment not subject to negotiation.

Should not human reason be strong enough to control destructive human "instincts"? It would be, said Freud in *Future of an Illusion*, if reason was not weakened by the teachings of religion and by credulous belief in devils and angels, heaven and hell. People brought up free of such "illusions" could admit to human helplessness. They would not think they were the center of the universe and God's special love children. As a result, they would put their liberated energy to work on human problems. Instead of thinking they could pray, bribe, cajole a higher power like a child begging parents for a gift, they would get to work devising ways to improve human life. Is there then to be no authoritative code of morality, no obligatory guide for behavior? But, responded Freud, historically religion has proven completely deficient as moral guidance. Focused on individual salvation, it does nothing to reduce selfishness. The relativity of religious ethics creates constant bloody even genocidal conflict. Inevitably established religions end by giving divine authority to prohibitions that are useless or discriminatory. Nature is full of terrors, conceded Freud, but imagining an illusory Providence that turns evil into good does little to control or ease those terrors. By all evidence, a less rigid fallible pragmatism would serve better to promote human happiness. Without the monotheist's illusion of privilege as the "love-child" of God whose needs will be met by a heavenly father, energy could be released and put to work finding ways to improve human life. Writing in the early part of the twentieth century, Freud was not optimistic. As he saw it, so strong was the grip of theism on the Western mind that progress in redressing human ills could be imagined only in a distant future when belief in God has finally disappeared.

* * *

For many popular writers on science in the last decades of the twentieth century, that time had come. In *The Moral Landscape*, neuroscientist Sam Harris argued that, given it must now be conceded that the sole correct definition of "good" is the "well being of conscious creatures," morality can be solidly based on findings in neurobiology and sociology and rise to the level of a science based on evidence and reasoning rather than prejudice and superstition. But are science and a concern for social good a sufficient basis for morality? For the defenders of religion, the answer was still no. Without divine sanction, without assured rewards for goodness or punishment for evildoing, where was the incentive for doing good? Without the expectation of life after death, what reason is there to give up one's own pleasure or happiness for the good of others? Without an omniscient God who knows all, vice can go undetected. Without divine Providence, a virtuous man who suffers in this life looks forward to no final bliss. The very fact, argued Kant, that people can and do on occasion act out of moral obligation against their own desires and instincts requires the rational supposition of judgment after death.

On this view, God is necessary as the active force that—like Lewis's ambitious artist, stern trainer, improving father, and demanding husband—presses down on believers to make them good. Without that pressure, a person would only try to look good before others, putting a positive gloss on actions however selfishly motivated. So it seemed to Thomas More, imagining the possibility of a rationally ordered communal society in his 1516 *Utopia*. Any virtuous society, More concluded, has to retain at least the core tenets of monotheist belief that a higher power exists, the soul is immortal, and after death individuals will be punished or rewarded. Otherwise, no one could be trusted to hold public office. No one would refrain from doing evil if he thought he would go undetected. Centuries later, Jean-Jacques Rousseau, imagining an ideal participatory democracy, agreed. In a free republic where the people are sovereign, some version of state religion has to remain to motivate more than token compliance with the law. Even the sternly anticlerical Robiespierre, presiding over the French Revolution's reign of terror, in the end had recourse to the worship of a "Supreme Being" to motivate citizens to moderate selfish desires and act for the good of the nation.

In this regard, biblical ethics is often claimed by historians of religion and religious scholars to constitute the great moral advance. As the argument goes, the good life envisioned by "pagan" philosophers like Epicurus and the Stoics focused on the happiness of individual mortal selves. At death either nothing survived or the individual soul dissolved back into the eternal soul stuff of the universe. In neither case was there an incentive to care about others. Epicurus argued that friendship is part of a happy life, the Stoics that moderation of hostile feelings was a way to avoid pain and disappointment; in both cases, the emphasis was on an individual's inner state of happiness. This "eudaimonism"—the view that a life well lived is a happy life—had to be replaced by Hebrew and Christian religion before there could be a true morality of justice and equal rights. Only with biblical ethics could selfishness be overcome by an altruistic morality of concern for the poor and respect for human rights.[4]

In *Leviticus*'s dense thicket of proscriptions and prescriptions relating to sacrifice, ritual purity, sexual hygiene, profitable management of property, and legal penalties for theft are buried a few injunctions to respect the rights of others: God's people should not curse the deaf or trip up the blind (19:14). They should not harvest to the very borders of their fields and not gather up gleanings so as to leave some pickings for the poor or in one case a "stranger" (19:9–10; 23:22). Several passages call for proper treatment of "neighbors"—do not lay with a neighbor's wife, do not rob or oppress him, love him as you do yourself, where "neighbors" clearly indicates fellow Hebrews. In at least one case, Hebrews are told to treat a "sojourner" or traveler passing through their community as a neighbor, because the Hebrews themselves were once sojourners in Egypt (19:33–34). Otherwise the overwhelming force of *Leviticus*'s rules and regulations is to distinguish the purity of Jewish practice from the idolatry, harlotry, and the evil abominations of other peoples inhabiting the lands promised to the Israelites by their God. The passage in *Leviticus* 19:18 cited by Jesus as the second great commandment, "Love your neighbor as yourself," read in context relates not to human beings at large but to "the sons of your own people," a stricture necessary in mediating the factional quarrels that continually divided the Hebrew community.[5]

From Hebrew prophets came periodic calls for charity. Amos cried out against the luxurious lifestyle of the newly prosperous Israelite kingdom under Jeroboam II and warned of the growing threat of Assyria. He singled out for special condemnation the women of Bashan, rich Samarians who "oppress the poor and call to their husbands to bring them drink" (*Amos* 4:1). He cried out against those who built houses of stone and planted beautiful vineyards while the poor were not being fed (5:11). A generation later, with the North annexed by Assyria and Judah an embattled tribute state, Isaiah called on God's people to redress oppression, defend orphans, and protect widows. At considerably more length, he condemned the "filth" of "daughters of Zion" who celebrate with music and wine. In the end, said Isaiah, Yahweh himself would have "no compassion for Israel's orphans and widows" but would destroy the weak along with the rich (9:17).

In any community involved in constant conflict with neighbors, provision has be made for widows and orphans, if for no other reason than that warriors are unlikely to be willing to fight if their wives and children will not be cared for if they die in battle. It is hard, however, to see a call for justice for the poor and defenseless as a major biblical theme compared to fears of extinction and dreams of entitlement of a small embattled people, encircled by imperial powers and at the mercy of powerful neighbors. It was customary to allow the poor to gather bits of fallen grain in harvested fields. Worshippers were expected to contribute something in the poor box in the temple and provide protection for women widowed by war and orphaned children. Much more insistent and unrelenting than these reminders is diatribe condemning the abomination and filth of peoples considered to be Israel's rivals and enemies. Jesus's insistence that followers "turn the other cheek" and "love their enemies" can be interpreted as cautionary. To oppose a Roman tax collector or centurion might mean imprisonment or a fine that would interrupt a believer's preparations for the final days when Israel's enemies would be punished and the elect would go to heaven.

Why must morality begin with loving God? Why would humans only care for their neighbors if they are first made to believe they are loved by a higher power? The standard Christian response is Augustine's: humans are selfish creatures, concentrated on their own happiness as evidenced by Adam's disobedience in Eden. Given this taint, and its subsequent punishment, more than "human nature" will always be required to make humans good, more even than religious law and fear of retribution. Only if humans can be made to feel God's love and love God back can some of that love radiate back onto fellow humans. On this argument, none of the various formulations of secular or philosophical ethics—acting on universal principle or bringing about the "greatest happiness of the greatest number"—provides sufficient motivation to make a person care for the well-being of others as much as he cares for himself.[6]

Again, monotheists and secular ethicists share common ground. Humans are naturally self-interested. They are born as detached individuals, looking out on an external world that may or may not meet their selfish needs or satisfy their cravings. The achievement of virtue requires the difficult if not impossible task of overcoming primal selfishness by recognizing the rights of others and coming to care for others' happiness as well as for one's own. The conflict between monotheists and secular ethicists is over how, given this essential human selfishness, that ethical goal can be achieved, whether by law—divine commandments, laws of reason, or manmade legislation—or, in the case of Christians, by loving the commanding God of biblical tradition. Only in this way can moral balance be established between the essentially conflicting demands of freedom and equality, individual and public good, selfishness and altruism. If there is to be some degree of "distributive" justice, there must be a point at which individuals are willing to give up at least some of the pursuit of their own individual happiness for the good of others. Either God must dictate a rule by which the rich are ordered to leave a remnant of harvest for the poor, or a rational basis must be found for some attention to the needs of others. In each case, compromise must be made between an individual's desire to do as he wishes without concern for others and a moral obligation to do something for others. The task of religious ethics and secular ethics is fundamentally the same: monitoring and regulating what is seen as a primal opposition between individual and public interest.

A similar divide between self and other informs debate between secular social ethicists over how to reconcile individual "freedom" with the "inequality" that inevitably results when competing individuals pursue selfish interests unrestrained by obligations to others. In some versions, a principle of "justice as fairness" proscribes formulas of adjustment. In a free market, individuals pursue their own self-interest, but the inequality that results can only be morally justified if the acquiring of pleasure by privileged individuals benefits to some degree those least advantaged. Inequities are fair when the disadvantaged would have been worse off without the inequities.[7] In other variations, the greater well-being of some individuals must be proven to be for some "public" good. To that end, the freedom of competing individuals must sometimes be curtailed so as to ensure the "dignity" of all citizens, at the same time as certain basic individual liberties of freedom of speech, the vote, and the right to acquire property are retained for all.[8]

Diotima's moral teaching does not begin from the same original position. The achievement of virtue, as she describes it, does not have to overcome a primal alienation between self and others, any more than it has to mediate a Platonic conflict between bodily instinct and rational soul. In the *Phaedrus*, a spirit soul is imagined trapped inside an alien body, fighting against urges that drive that body to shameful acts. Imprisoned in that pleasure-seeking body, the rational soul looks out on other bodies from a position of primal isolation. In the modern psychoanalytic version, the bad horse of the *Phaedrus* is the libido, a biological urge seated deep in the human psyche, dangerous to self and others when not under the control of a rational calculating ego supervised by a parental superego. In postmodern versions, discursive reason creates alienated identities that are periodically troubled by irruptions from unquantifiable inner drives and impulses. Diotima's humans are not Platonic souls in bodies, divided Freudian psyches or constructed selves with roles and identities fixed within social or linguistic structures.[9] Like Phaedra, a woman may indeed turn against herself in self-hate. Like Lucretius, a man tormented by jealousy and loss may come to have contempt for the opposite sex. A person may indeed become an automaton, playing "language games," acting out roles scripted by social norms, going through the motions of life without original response to people or things. Whole societies may exist in which human behavior is scripted by caste, birth, or totalitarian political regimes. But these do not represent an original human condition. In any such society, there is always the possibility of that shrinking away from ugliness and quickening response to beauty that generates affiliation, common cause, and of new ways of thinking and acting. On this view, no matter how repressive are existing social norms, humans are inherently conscious physical beings naturally "pregnant" and responsive to other physical beings.

There is no radical separation between soul and body, nor is there any primal separation between the needs of self and the needs of others. From infancy the need to eat is the need to be fed. Bodily sensations are intimately bound up with responses to people and things outside oneself. Babies respond to faces almost as soon as they learn to suck. The primal human condition is nexus and interaction as human needs are satisfied in combination with others who are not simply bodies but responding sentient beings. Humans are not isolated souls looking out from the bars of a bodily prison onto other bodies that may or may not have souls inside them, but are from infancy linked to others by response and interaction. It is not the difficult, perhaps impossible, achievement of altruism that is virtue as Diotima sees it, but rather the expansion of narrowly focused relationships to wider networks of productive affiliation.

* * *

Since Plato, moral theory has been divided into two broad categories: moral relativism—what is good depends on subjective personal feeling or variable community standards—and moral absolutism in which one true standard of virtue and goodness is prescribed for all situations and times. Diotimean ethics falls into neither category. There is no transcendent model or text by which to distinguish "good" from "bad" beauty, no grounding for the claim that one can define and teach what beauty always is. There is no attempt in her teaching to find the necessary truth from which to infer an always reliable moral judgment. If truth is found in justice, natural law, God's commands, or utility, that truth can always be called into question once it takes form in a particular practice or outcome. As Socrates points out to the sophist Hippias, for any supposed absolute beauty there are circumstances in which that beauty may be ugly rather than beautiful. Well-made speeches can deceive the public and lead to rash and unwinable wars. What is useful may be useful for a bad purpose. What looks beautiful to eat can make you sick. But neither is individual preference the standard of virtue. Individuals and groups of individuals may refuse to see ugliness; they may be blind to beauty and destroy it. It is here that the interactive matrix of Diotimean virtue is so important. The virtuous person does not fall in love with a particular institution, ritual, or belief, but with persons with whom she or he makes common cause in "giving birth" to beautiful results. It is always possible, and even perhaps inevitable, that the beauty of human arrangements may be corrupted. For this reason, there is no one eternally beautiful template for an ideal society or way of life. Only women and men responsive to ugliness and emergent beauty can become "begetters of virtue" and, in the spirit of Plutarch's Isis, restore harmony to human affairs.

Is this simply to displace the relativism/absolutism dilemma from goodness to beauty? Should not the failure of eighteenth-century attempts to systematize aesthetics prove that the appreciation of beauty is simply the enjoyment of a certain kind of pleasure, the source of which may differ from person to person? Here is where the transcendence of the revelation that makes the initiate in "matters of love" a begetter of true virtue and friend to the gods must be taken seriously. After the progressive enlargement of sensitivity and cooperative creation of beauty in all its forms comes that sudden grasp of beauty as a universal and eternal transcendent dimension of physical existence and consciousness. It is the very lack of objectivity, of being definable as an identifiable object, that gives the final revelation both its divinity and its moral force. The effect is radically different in kind from knowledge that a certain set of commandments, actions, social structures, outcomes, or systems of principles define true virtue. Beauty itself is something more than a fashion in aesthetics; it is the universal sense of the attractive power of potential good beyond any present reality.

In this sense, the moral force of divinity is not the heavy weight of authority pressing down on innately selfish human will; it is the pull of aspiration to more beautiful and lasting ways of living. Diotima's final vision of divinity requires the matrix of interpersonal creative relations between lovers, families, and members of a community. It requires long experience with frustration and loss, as well as moments of happiness in creating and contemplating beauty in nature, music, social arrangements, or science. What the initiate into the mysteries of love can aspire to is not a grasp of the transcendence of one or more supernatural persons or principles, but transcendence itself as an ever-present and eternal aspect of nature and human life. Daemonic spirituality is necessary for the achievement of virtue, not because human "will" is tainted or fallen, not because an all-knowing God can detect inner evil thoughts, and not because there must be a reward for virtue in an afterlife. It is necessary because even the most rigorously scientific of initiatives to reduce suffering and promote well-being must begin with "the lure" of the never disappearing possibility of evolving harmonious ways of life. In animal life, some of this natural aspiration beyond the mortal self motivates the choice of mates, the care of offspring, the herding together in groups. With humans comes a vast widening of erotic virtue and cooperative creativity in systems of governance, culture, art, and knowledge.

CHAPTER 11

The Problem of Evil

When those pregnant with unborn conceptions come near what is beautiful they are gracious and happy, over-flowing with begetting and creating, but when they come near to what is ugly, angry and tense, rejected and wound tight, they give forth nothing, but go bearing the heavy burden of unborn offspring.

—Symposium *206d3–9*

For theologians charged with making sense of monotheist doctrine, "evil" has always been problematic. If God is the creator of all that exists and at the same time perfectly good, why does evil exist in the world at all, let alone personified as demonic beings that can challenge even God? If the animal body is the site of sin, why was the body created? Why not soul creatures free of bodily urges? If the taint is not in the body but in human will, why is man, God's creature created in his image, so imperfect? And why do men on occasion murder and rape innocent victims? Can God like to see people suffer? Does he test them? Does he torment them simply to see how long they can hold on to their faith before they turn and curse their maker? Is God's power simply limited? Did God, as Leibnitz argued, create the best of all worlds that it was possible for him to create?

In monotheist debates surrounding the problem of evil, Paul's "pornea" and Augustine's "concupiscence" proved particularly challenging. Deeply embedded in natural life, necessary for reproduction, breaking out no matter attempts to repress it, sexual desire dominates Christian ethics. Is premarital intercourse immoral? Does contraception promote promiscuity? Is it a sin to desire someone of the same gender? Is it best, as Paul urged on the Corinthian congregation, to remain chaste and devote oneself to God? At the same time, as Paul freely acknowledged, sexual desire is powerful and when repressed leads to the evils of prostitution, pornography, child abuse, and rape.

In the first decades of the twentieth century, under the influence of Freudian psychology, expressions like "sex drive" and "libido" came into use to refer to a new secular psychoanalytic version of concupiscence.[1] Freud based his account of human psychological development not on a myth of original sin but on an individual's attempts to manage his own pleasurable and painful sensations. Little by little a human infant begins to distinguish between the internal and external causes of his pleasure and pain. Soon, as a rational self with a grip on the reality of those others "out there" on whom so much of his pleasure depends, he begins to practice what Freud called "the economics of the pleasure principle." Controlling and manipulating aspects of external reality, he works to increase his pleasure and avoid pain in a world often indifferent or opposed to his needs and desires. To make the task more difficult, pleasures are fleeting. Sensation dulls, interest wanes. Long-delayed gratification brings a moment of bliss, which is as quickly gone. This is especially true of sexual pleasure. Orgasm is intense but over quickly, momentarily gratifying, but offering little in terms of steady contentment. It can seem, lamented Freud, that far from showing any evidence of divine Providence the cosmos was not designed to make man happy.

> [The pleasure principle] dominates the operation of mental apparatus from the start. There can be no doubt about its efficacy, and yet its program is at loggerheads with the whole world, with the macrocosm as much as with the microcosm. There is no possibility at all of it being carried through; all the regulations of the universe run counter to it. One feels inclined to say that the intention that man should be "happy" is not included in the plan of "Creation." What we call happiness in the strictest sense comes from the (preferably sudden) satisfaction of needs, which have been dammed up to a high degree, and it is from its nature only possible as an episodic phenomenon. When any situation that is desired by the pleasure principle is prolonged, it only produces a felling of mild contentment. (*Civilization and Its Discontents* 23)

Frustration and anger inevitably result, leading to societal and parental restraints and the eventual development of an internalized authoritarian "superego" that rules many pleasures wrong.

In life ruled by Freud's "pleasure principle," "sex"—like Augustinian concupiscence—provides the most vivid and problematic of pleasures, so vivid that in civilized societies the "sex drive" has to be repressed to ensure that time and energy are left for work, hygiene, and the care of children. Not only does the urge for sexual gratification have to be deferred, it has to be made to take certain forms and attach to appropriate objects. Again the key to a degree of comfort and mental health, said Freud, is judicious regulation, decisions as to "how much real satisfaction a person can expect to get from the external world" (*Civilization* 30).[2] For Freud, "pair-bonding" in marriage is one of these "economies," a compromise in which a man sacrifices some of the pleasure in sexual novelty in return for having a convenient sexual object near and available. Why not make pair-bonding the center of one's life? Here, said Freud, there is another danger. Putting too much energy into a marriage leaves the monogamous "erotic man" dangerously dependent on one object of pleasure that might leave him or die. Weighing profit and loss, the amount of pleasure to be gained from regular sexual intercourse may or may not outweigh the uncertainty of controlling the object on which that pleasure depends. The frustration and tension that results can lead to incapacitating mental symptoms as well as to behavior destructive to self and others. Adding to the difficulty is the fact that repression of libido whether in the form of external legal punishment or an internalized sense of guilt only makes aggression stronger, forcing it inward against the self or outward against others in acts requiring ever harsher repression and leading to ever more violent aggression.

In the early decades of the twentieth century, with anti-Semitic violence on the rise in Germany and world war pending, aggression was on Freud's mind. Early on he noted deep currents of hostility and anger in his male patients as well as in himself, a fact that he tentatively attributed to the unsatisfying nature of pleasure, societal and parental repression of sexuality, and a son's Oedipal resentment of his father's authority. By 1920, writing *Beyond the Pleasure Principle* in the wake of the brutality and carnage of World War I, he was ready to posit along with the economics of the pleasure principle, a related "death wish," a psychic drive aimed not at pleasurable survival but the annihilation of self and others. Aggression, he began to think, could not be blamed simply on the fleetingness of pleasure or analyzed as a byproduct of unresolved complexes but must be a demonic evil inherent in human nature. Ten years later, writing *Civilization and its Discontents* in the midst of the hatred and violence that would eventually lead to fascism and another war, he was more and more convinced that along with the pleasure principle there is in the human psyche "from the start" a rival drive to destruction. He pondered. Is there an instinct to harm others that rules out any possibility of brotherly love? Is hatred a drive so powerful that, no matter how vigorously "aim-inhibited" caring for others is promoted, no matter what harsh punishments for violent crime are imposed, civilization can never touch "the more cautious and refined manifestations of human aggressiveness" (*Civilization* 59). Aggression, decided Freud, must be innate, a deeply implanted death wish aimed at self and others and somehow linked with eros. Take away property men fight over women or honor. Establish territorial boundaries and nations deflect aggression onto hatred of minority groups.

Over and over, as the threat of war came closer, Freud returned to the puzzle of what he saw as a "destructive instinct allied with eros" (67), at the same time as he continued to come back to the "impossible" idea of eros as a source of communal harmony.

> I may now add that civilization is a process in service of Eros, whose purpose is to combine single human individuals, and after that families, and then races, peoples and nations into one great unity, the unity of mankind. Why this has to happen we do not know; the work of Eros is precisely this. These collections of men are to be libidinally bound to one another...but man's aggressive instinct, the hostility of each against all and of all against each, opposes this program of civilization. (*Civilization* 69)

At various points in *Civilization and its Discontents*, Freud came close to acknowledging the positive social value of eros. At the end of a list of possible ways to ease the pains of physical existence—drugs, yoga, sublimation, fantasy, retreat from the world—he notes that he has left out "love," love that not only "has given us our most intense experience of an overwhelming sensation of pleasure" but "has furnished us with a pattern for our search for happiness" (*Civilization* 29). Even though it can result in loss, might not "pair-bonding" be a beginning point for wider affiliations? But, Freud pondered, this could only be true if love is the "absurd" Christian agape, love of others regardless of any merit on their part.[3] No, he concluded, there is irresolvable conflict between eros and the higher accomplishments of civilization. Once two individuals become one in love, eros has accomplished its purpose; there is no further reach of erotic desire. Still, he mused, if sex were simply the urge for pleasure, why does pair-bonding occur at all? Why would one object be any more important than another? For Freud, there was no way out of the quandary. Committed to the idea of the psyche as a complex of interlocking internal instincts and the external world as a source of manipulatable mechanisms for serving those instincts, he can see no way to move from sexual love to wider interests in science, art, or civic and must conclude that painful repression of eros is necessary if the higher accomplishments of civilization are to be possible.[4]

At the same time, Freud confessed that there was much that could not be explained about eros within the framework of psychoanalytic theory. If a pleasure-seeking ego trying and sometimes failing to control an unpredictable external world is the natural human condition, falling in love, as an experience in which the boundaries between self and other fall away, would seem to constitute the anomaly of a "normal" exception. But stronger than this apparent contradiction was Freud's certainty that men who live only to love must give up any ambition to be scientists or artists and, clinging to their wives, must remain housebound and domesticated. If the model for Socrates's mythic painting in the *Phaedrus* is the conflicted pederastic lover of boys, the model for Freud's "erotic" man is the bourgeois European husband. Taught by religious authorities that lust is a sin but tacitly given license as a man to satisfy lustful cravings as long as respectability was maintained, such a man was expected to marry and produce heirs. Dependent on one female object for sexual pleasure and subject to the needs and demands of domestic life, it had to seem a bad bargain. As Freud noted, a spasm of genital pleasure, however licit, leads to little in "steady contentment." Making matters worse was the veil of social propriety covering over what was seen as, and often was, the nastiness of "sex." Children, both male and female, were physically and sexually abused by male relatives, schoolteachers, and priests. Forced to keep silent out of shame and family loyalty, the anger and shame of victims was carried into adult life. Once married a husband's duty was to perform sexually, but as with Paul's *pornea* and pederastic "*aphrodiziasing*," Freudian libido had none of the urgency of transcending mortality and creating lasting good. The desire of women to hold together a family and not be separated from their children was regressive. To achieve anything in the world of men, a husband had to escape female demands. As Freud put it:

> Women soon come into opposition to civilization and display their retarding and restraining influence—those very women who, in the beginning, laid the foundation of civilization by the claims of their love. Women represent the interests of the family and of sexual life. The work of civilization has become increasingly the business of men, it confronts them with ever more difficult tasks and compels them to carry out instinctual sublimations of which women are little capable. (*Civilization* 50)

Even if a man's love might extend to family and friends, Freud argued, any attempt to go further would remain subject to the "retarding and restraining influence of women."

Late in life, Freud confessed that there was much about both women and female sexuality that psychoanalysis had not touched. The sexual life of women, he lamented in "The Question of Lay Analysis" (1926), is a "dark continent that we analysts know little about." He acknowledged that he had not been able to find any clear parallel between the sexual development that he mapped out from his own and his male patients' experience and women's experience. Somehow, he speculated, the cause must lie in the shadowy stages of early pre-Oedipal development and a girl's relation with her mother, a mystery he compared to the inscrutability of the pre-Hellenic pre-Christian civilization that archeologists were unearthing beneath the monuments of classical Greece.

> Our insight into this early pre-Oedipal phase in the girls' development comes to us as a surprise comparable in another field with the effect of the discovery of the Minoan-Mycenaean civilization behind that of Greece. Everything connected with these first Mother attachments has in analyses seemed to me so elusive, lost in a past so dim and shadowy, so hard to resuscitate, that it seemed as if it had undergone some especially inexorable repression. ("Female Sexuality" 1931)[5]

But musing on the growing hatred of Jews and political tensions in Europe, Freud soon lost interest in any forgotten Mediterranean world of meaning and experience and turned to speculative revision of the history of Judaism in *Totem and Taboo* and *Moses and Monotheism.*

* * *

Diotima's lessons in love suggest a different way to understand the evil of violence and aggression. Rather than a craving for a spasm of genital pleasure, Eros, generated out of need and cleverness, is aroused in the presence of beauty. But when confronted with the ugliness of sexual abuse, confinement to domesticity, hypocrisy, threats of religious damnation, a person, no matter how teeming with life, draws back, "angry and tense, rejected and wound tight." Not only are spirits dampened, but, recoiling painfully into his or herself, such a person goes away with a heavy burden of unborn aspirations. Repeated over and over, such experiences can result in a permanent state of hostility either against oneself or projected out against others. Especially this is true when the ugliness confronted is abject poverty or racial hatred that denies to individuals resources with which to achieve what they desire. In such situations, a person can come to repress and hate his bodily existence, and by extension the physical being of others. The urgency of wanting to surpass one's mortal existence finds no outlet; frustration and anger turn against others. Sex becomes a violent act, an assertion of mastery, a taking of pleasure at someone else's expense, including the kind of violent retribution on women's bodies described by the prophets Hosea and Jeremiah. Action can become a pure assertion of will, unhinged from any ultimate purpose or meaning.

When religious and sectarian differences sharpen that hostility, what can result is the breakdown in the fabric of human society that Plutarch alluded to as the chaos of Typhon, that Empedocles evoked in his *Purifications*, and that Freud and others experienced in the Great War and the rise of fascism. Demonization of enemies, murderous reprisal, suicide bombings tear away any sense of community. Fields and houses are destroyed and children lost. Militarism, ethnic cleansing, terror, and death are the rule. Any hope of Providence or a "plan of Creation," as Freud put it, is a cruel joke. Even when a degree of civic order is restored, there can be no certainty that good will prevail. Abandoned illegitimate boys like Theseus may become bullies and adventurers, and later repressive heads of state. Women like Phaedra maneuvered into loveless marriages may end by destroying themselves and others. No righteous Christian standard in battle, no superior force, no treaty between heads of state can redress or repair the damage done. All that can begin to restore peace and order are the signs of life Plutarch alluded to as the stirring of Isis, as erotic affiliations revive and new social forms come into being with the rebuilding of ruined homes, finding of lost children, planting of garden plots, reunion of lovers, and formation of citizen councils. In this work the generative urge of eros is not the downfall of civilization but its salvation.

Contemporary moralists often follow Paul and Freud in denouncing eros as a barrier to social good. For the achievement of virtue, argued Martha Nussbaum in *Upheavals of Thought*, eros must be "emended."[6] People must get free of the joys of sexual bonding. They must come to feel "general social compassion" within "reasonable limits" based on a "just determination of the seriousness of the human condition in question." They must be taught to make a judicious assessment of human need. They must come to have some sense of "reciprocity." They must treat people as individuals with rights. They must give to charity. They must recognize sensible limited obligations to others instead of focusing solely only on their own libidinous needs and desires. She cites Rawls quoting David Hume: Social justice must start from "the perception that we are in a situation of competition for scarce resources that we badly need" (498). Eros must be curtailed, said Nussbaum, so that there is some degree of response to need in others and so that the individual choices of diverse citizens are respected and social order is maintained.

The psychological, ethical, and social framework of a Diotimean approach to evil begins from a different understanding of the human condition. Tithing, charitable giving, legal provision of a minimal social safety net for the poor, all are loveless acts requiring no actual involvement with the needs and desires of disadvantaged communities. The motive force behind good works in Diotima's teaching is not judicious consideration of the least benefits that others have a right to expect from those better off, but response to both need and possibility. If behind Freud's distrust of eros was the troubled bourgeois family, behind Nussbaum's emendation of eros is the democratic capitalist state founded on an extension of the "economics of the pleasure principle." If individuals compete to further their own interests, there will be more total pleasure in the form of wealth. In such a state, some redress of suffering is expedient. Those at the bottom of the economic ladder must accrue some benefit from the system, if only because without access to food or some form of health care they may eventually rebel. In such situations, compassion and charity may relieve some of the suffering but cannot alter the alienated social conditions that cause that suffering.

The "begetting of virtue" of Diotima's "friend of the gods" takes a different course. She might see beneath the surface ugliness of a degraded slum small buds of beauty of spirit with generative possibilities. Virtuous social action would not take the form of relief or charity, but rather community development and activism. Nor would there be a predetermined course for that development. Instead as social activist she draws on the values of the disadvantaged community and works with its members to define goals and priorities. The model for ethics is neither charity nor a reciprocity of private interest (you promote my interests, I'll promote yours) but seeing, responding to, and allying oneself with the beginnings of enlivening spirit in individuals and communities worn down by ugliness and need. A virtuous reformer might join and help to organize unions of exploited workers. She might work with groups of local farmers to market their products and receive the full value of their labor. She might go to town meetings to prevent harm to the environment.

It is no accident that the Aphrodite associated with Pausanias's plebian eros, was Aphrodite "of the people." Virtue, as described by Diotima, is not for a privileged educated elite who can afford to calculate and redress some of the pain of inequality their privilege requires, but is suited to ordinary men and women. Virtue is not respect for rationally parceled-out "rights" but interaction and common cause in the interest of better, more beautiful forms of life. It is working with others to eliminate the ugliness at the heart of the despair and frustration that leads to vice and aggression. Mentors, teachers, parents help young people to see beauty and propagate more. Scientists enlist students in the search for ways to cure the painful ugliness of disease. Legislators in touch with the values and interests of those they represent suggest new social initiatives. What results are not remedial measures to reduce some of the suffering and violence that a particular social system causes but parties, institutions, and organizations that empower groups of people to create new forms of life. Eros is the driving force, not as love of an abstract idea of justice but as productive relations with others in work, the arts, scholarship, and community organizing.

* * *

Diotima's final vision is of empowered transcendence, the knowledge that human life can change and grow better with human initiative and response to visible beauty that energizes erotic energy. Similar was the "new language of transcendence" called for by feminist theologian Mary Daly in *Beyond God the Father.* Liberation movements, said Daly, cannot do without religious transcendence, otherwise a cause becomes an "idol" with all energy focused on one goal such as the right to vote or run for office. Instead, wrote Daly, "it is the creative potential itself in human beings that is the image of God," as opposed to false gods used to explain away evil, offer an afterlife as reward for obedience, or give authority to unfair laws (29). What Diotima's lessons in love make clear is the importance of human intercourse in realizing that transcendence. For social movements to bear fruit, supporting any electoral process must be generative networks of affiliation that allow individuals to conceive new possibilities for change and devise methods to bring them about. Essential in this process is the "conversing together" which allows Diotima's lovers to bear and give birth to conceptions "they have held for some time" (*Sym.* 209c3–4).

In the feminist spirituality movement of the last decades of the twentieth century, in which Daly was a prominent figure, different groups of women with long-held convictions—straight white middle-class women, lesbians, women of color—came together in passionate discussion. As a result, feminist theory and feminist action took new directions generating movements in language reform, gay politics, critical theory, and religion. Energizing and sustaining these encounters and initiatives was not compassion for victims more oppressed than oneself or the conviction that everyone black or white, gay or straight should enjoy certain minimum rights, both of which could have been taken as demeaning by women of color and those stigmatized because of sexual preference. In gay women, white women saw the beauty of women creating new kinds of romantic partnership and child-rearing. In women of color, white women saw images of strength and resilience compared to the ugliness of white privilege. As a result, ideas of liberation expanded, surpassing the vote and the right to equal employment to envision new cultural and societal norms.

Another obvious application is in environmental ethics. Any line of argument from biblical ethics to environmental protection or concern for the evils of climate change has proven tenuous. One passage in the *Genesis* creation story seems to enjoin stewardship and care of nature; much more influential have been passages in which God gives nature to man to dominate and use as he pleases. The subsequent biblical message that floods, earthquakes, famines, and epidemics are righteous punishment for sin and idolatry invites complacency if not the persecution of victims. The expectation of a final apocalyptic end to nature downgrades the value of life on earth. In contrast, love of the world and response to natural beauty provides the pull of aspiration to resuscitate nature's power of regeneration. The continuity between humans and other living beings assumed in the teaching of Diotima suggests new forms of partnership with nature. The goal is not return to an Edenic wilderness whose God-given balance was disturbed by original sin, nor is it the efficient use of resources to produce more wealth. Instead, as in human love the impulse is to creative interaction between human cultures and the nonhuman cultures of other living beings. The ugliness of strip mining, dead coral reefs, oil slick, all are repellant and give rise to injunctions preventing further damage. Even more important in motivating renewal would be the ability to see beauty in nature and the resulting desire to work with plants, animals, and climate to restore and tend that beauty.[7]

An evil even more challenging are sectarian hatreds that turn whole regions into murderous violent chaos, which can end only in mutual extermination with houses turned to rubble, children massacred, fields and houses torn apart, all livelihood gone beyond any lifeline of international aid. Given the mixing and mingling of peoples and traditions that modern means of travel make possible, given artificial boundaries established by imperial powers, those of different religious groups inevitably live in close proximity. As Elizabeth Johnson puts it in *Quest for the Living God*:

> In our era of global communication, planet-wide travel, widespread immigration, and millions of refugees, religious traditions are jostling

> against each other as never before... However committed persons may be to their own faith, they are regularly confronted with people whose commitments offer a different claim about what is worth believing. (153)

When each sect claims that its own version of what God wants or commands is true for all people and for all time, differences become irreconcilable resulting in forced conversion, segregation, and religious and ethnic cleansing. Borders between separatist neighborhoods, enclaves, and regions become lethal sites of confrontation and armed conflict.

Johnson addresses the problem of interfaith conflict from a premise close to Diotima's "gods do not mingle with men." Given that the "incomprehensible mystery of God lies beyond all human control and understanding," human access to divinity "throughout history" has been by way of "little theophanies, signs, and revelations and events that invite us into relationship" (*Quest* 161). Given this daemonic diversity, the answer to interfaith conflict is not separation and differentiation but "dialogue," the coming together in earnest conversation and discourse that in Diotima's teaching is the beginning of the "work of love." Like initiation into the mysteries of love, interfaith dialogue, as Johnson describes it, is progressive. It may begin simply with sharing the ordinary habits of life, shopping in markets, attending the same schools, and living on the same streets. When those of different religions begin to find common cause and work together for safer streets or just laws, dialogue may move on further to matters of knowledge in the form of theological exchanges and study of other belief systems that results in deeper and richer understanding of one's own tradition and beliefs. Finally, in the spirit of Plutarch's advice to Clea, there may be actual participation in other forms of worship, a sharing of prayers, fasts, pilgrimages, or religious holidays and rituals. Johnson's metaphor of music—"If one has an ear attuned to the Spirit's presence, then one hears the music, even if it is being played in a different key"—echoes Diotima's teaching that the daemonic is "multiple and takes many forms" (*Quest* 174). No such result is possible without Diotima's "great daemon" eros, understood as the ability to see and respond to what makes beauty visible in other persons and practices. It can come as no surprise that often it is falling in love across a seemingly unbridgeable ethnic or religious divide that begins a dialogue that eventually restores peace and harmony.

CHAPTER 12

The Defeat of Death

Then, giving birth to true virtue and raising it up, this person will become a friend of the gods and, more than any other mortal being, immune to death.

—Symposium *212a7–10*

That mortal creatures die is a universal truth of organic life; they also reproduce and after death become the material for new life. The most dedicated of materialists agrees that the dead rise up in the form of plants and animals that live on plants. The most vehement of atheists agrees with Paul in *Corinthians* 15 that seeds buried in the ground take on new bodies as shafts of wheat. Paul went on to interpret these teachings at the heart of the mysteries of Demeter in a new way. It was not that the dead bodies of Christians would become the joyful "first fruits" of next year's harvest; they would literally stand up with new bodies ready for a second life in heaven. He based this expectation on what he had fervently come to believe. A dead man had bodily risen from the dead, and his faithful followers could do the same.

The idea of immortality as literal resurrection of the body came relatively late in Jewish thought. There was little hint of it in the pronouncements of biblical prophets for whom deliverance meant victory in battle, acquisition of territory, and population increase. But in troubled years in Jerusalem before and after Jesus's birth—the humiliating turmoil of a failed rebellion, warring factions, deified images of Roman emperors brought into the Jewish temple—hope dimmed that the ancient Kingdom of Israel would be literally and materially restored. Biblical prophets blamed the Israelites themselves for their defeats. They had not been diligent in worship and obedience; they had acted the whore and gone off to worship Astarte and Asherah. But repression of rival sects, centuries of self-castigation, rigorous observance of the Sabbath, and adherence to rules of ritual cleanliness had not restored the Kingdom. Most troubling, some of those most zealous and assiduous in obedience to God's law had died as martyrs. All, then, must have been no more than preparation, if not for the eventual restoration of a prosperous and powerful earthly Kingdom, for something grander, the destruction of the entire demon-infested natural world with its reign of evil emperors and the restoration of the Kingdom of Jerusalem in a new age. Visions attributed to Daniel, a Jewish seer at the court of Babylon, ornamented with images from classical and Canaanite myth, revealed a secret plan prophetic of the occupation and desecration of the Jewish temple, proving that these earthly defeats were no more than prelude to cosmic victory.

> At that time Michael, the great prince who protects your people, will arise. There will be a time of distress such as has not happened from the beginning of nations until then. But at that time your people—everyone whose name is found written in the book—will be delivered. Multitudes who sleep in the dust of the earth will awake: some to everlasting life, others to shame and everlasting contempt. Those who are wise will shine like the brightness of the heavens, and those who lead many to righteousness will shine like the stars forever and ever. (*Daniel* 12:1–3)

Instead of launching another failed rebellion, the faithful could wait for God's final plan to be carried out, with Israel's enemies destroyed, its martyrs risen from their graves, and a final judgment of the wicked and the good. Was this not the meaning of the promises made to the Jewish elite languishing in exile in Babylon? Had not Ezekiel described being brought to a valley of exposed human bones, where God said to him: Do you think I can reassemble those bones?[1] When Ezekiel hesitated, God proved his powers, telling Ezekiel, "Say to them, dry bones, hear the word of the Lord," and at the command the bones reassembled, grew flesh, and stood up as a vast army. Was the moral not clear? Israel's fallen warriors would similarly stand up at God's command and fight again (*Ezekiel* 37).

With the Jerusalem temple rebuilt under Roman rule, apocalyptic writers fleshed out the idea of bodily resurrection for an elite few. The writer of *II Maccabees* revised the historical account of the Maccabee uprising in *I Maccabees* to include a promise of resurrection on judgment day when the King of the Universe would return. The *Book of Enoch* added hallucinatory detail with visions of heaven and hell, demons and fallen angels, a coming Messiah, and a heavenly Kingdom with resurrection of the faithful and evil oppressors tortured and punished. *Jubilees* gave new meaning to *Genesis* and biblical history as prophetic of a coming "end of days." Although apocalypse and resurrection were rejected by the more sophisticated temple elite—wealthily Sadducees happy enough with life on earth—scribes, craftsmen, lawyers found the idea of a beatific afterlife reassuring.

After the final destruction of the Jewish temple by the Romans in 70 AD, the tradition continued. The prophetic books of *Ezra* foretold a day when earth would give up those who sleep in it (4 *Ezra* 7:32). The *Sibylline Oracles* promised that God would put together again bones and ashes of dead men so that a final judgment could take place (iv:180). Scholars worked to find continuity between supernatural resurrection and passages from Hebrew Scripture with *Isaiah* a popular choice:

> On this mountain he will destroy the shroud that enfolds all peoples, the sheet that covers all nations; he will swallow up death forever. The Sovereign Lord will wipe away the tears from all faces; he will remove his people's disgrace from all the earth. (25:7–8)
>
> Your dead will live; their bodies will rise. You who dwell in the dust, wake up and shout for joy. Your dew is like the dew of the morning; the earth will give birth to her dead. (26:19)

In context, images of rebirth in *Ezekiel* and *Isaiah* might be read as metaphors for the restoration of Jewish rule. Passages that follow Isaiah's prophecy of last days foretell the coming of a leader who will punish the Israel's enemies, reassemble the collective Jewish people, and restore the temple in Jerusalem.[2] Jewish "zealots" continued to give political meaning to this final victory when Jewish independence movements would "rise again" from the "ashes" of defeat and Israel would regain the promised land. The idea of supernatural resurrection was harder to make sense of. What kind of body was this that was to be reassembled from rotted flesh and dry bones? How could a decomposed corpse rise up from underground except as a ghoul or ghost? If it was not the same body that was reassembled, how could it be the same person?

For a sophisticated few versed in Greek philosophy, Plato's winged souls detached from their bodies suggested another way to think about immortality. Was Plato's winged psyche not the same soul "breathed" by God into the body of Adam in *Genesis*? And having been breathed into a human body, might it not escape that body at death? Given biblical promises of the restoration of material prosperity and power, many of Jewish heritage were slow to accept eternal reward without at least some of the benefits of earthly conquest. Many found it hard to be consoled by the immortality of a spiritual essence that might dissolve away into the eternal soul stuff of the universe or, if it remained intact, could enjoy none of the joys of earthly existence. What resulted was an indistinct mix of two conflicting ideas of immortality, as survival both spiritual and somehow bodily.

As Jewish scholars and scribes debated the problem, Paul and his Christian followers relied on faith. Conceptual problems receded before what they took as proof of resurrection. Jesus had risen from the dead and promised that others could do the same. Paul admonishes skeptical Corinthians:

> But if it is preached that Christ has been raised from the dead, how can some of you say that there is no resurrection of the dead? If there is no resurrection of the dead, then not even Christ has been raised. And if Christ has not been raised, our preaching is useless and so is your faith. More than that, we are then found to be false witnesses about God, for we have testified about God that he raised Christ from the dead. But he did not raise him if in fact the dead are not raised. For if the dead are not raised, then Christ has not been raised either. And if Christ has not been raised, your faith is futile; you are still in your sins. Then those also who have fallen asleep in Christ are lost. If it is only for this life we have hope in Christ, we are of all people the most to be pitied. (*I Corinthians* 15:12–19)

If what was to be reborn could not be the same physical body, it would have to be a new spiritual body, "sown as a physical body, raised as a spiritual body" (*I Corinthians* 15:44).[3]

No issue, with the exception of the problem of evil, would exercise Christian theologians in the centuries to come as much as making sense of personal immortality as the primary benefit of Christian faith. In the afterlife that is to be the reward for faith, a Christian has to be judged as an individual. But once a soul is liberated from a physical body, it is hard to see how it can be differentiated as having an individual history, and if the soul cannot keep its identity, it cannot be judged. Nor could a Christian be judged as a soul that has gone through a series of bodies as in Plato. And if the individual cannot keep its identity and be judged, where is the reward for faith in Jesus or the incentive to follow Jesus's teaching in a world in which "evil" doers flourish and good people suffer?

* * *

Mark Johnston striving to make sense of a noncredulous naturalized Christianity puts the question succinctly in *Surviving Death*, sequel to *Saving God*. "Is it possible to ransom any of the genuinely salvific ideas found in the major religions from their supernaturalist captivity and what price do we have to pay for the ransom?" (*Surviving Death* 16).[4] At stake, said Johnston, is the survival and even the very possibility of goodness. Some sort of immortality is necessary if goodness is to be protected and rewarded. At the same time, in a scientific age the price cannot be credulity. Since there is no rational or scientific way to make sense of any identifiable stable soul entity or self-persisting after death, protection of and reward for goodness has to somehow come in the form of good things for other humans who live on after us. But here, argues Johnston, human selfishness stands in the way. Given that human experience is inherently framed in terms of a "me in here" with my interests, and others "out there" with their interests, it is hard to see how the good of others after we are gone can be of much interest to us.

There are, Johnston acknowledged, "natural" this-life advantages for individuals who care about the interests of others. Caring for others usually results in better relationships. Caring persons have less need to maintain fictional defenses and construct fantasies. They have a better grip on reality and an easier death. But "salvation" must constitute more than a good chance of having a happy life. Goodness must result "generically," in and of itself, in some sort of "reincarnation" or "embodiment." One solution might be Christian agape, especially if reinforced by the Buddhist tenet that there is nothing in the individual self worth caring about. Again, however, for Johnston the assumption of inherent self-interest makes reward in this form less than consoling. Given that a person has interests distinct from and in conflict with the interests of others, given that a person has a built-in tendency to give preference to his or her own interests, "loving one's neighbor as oneself," to say nothing of loving all human persons as oneself, seems all but impossible.

The difficult route to immortality finally mapped out by Johnston reverses Diotima's work of love. If, instead of seeing and responding to the beauty of others, we extend our own self-interest by giving preferential treatment to family members, then to our own tribe or ethnicity, and possibly to our nation in a kind of herd-like "species consciousness," a "thought may break through." Given that our preference for the interests of these extensions of ourselves does not depend on merit any more than does our preference for our own interests, we might rationally infer that the interests of others must be as important as our own. If family members are persons valuable like ourselves, then so is any person to be so valued. Having made this inference, argues Johnston, if a person happens to have sufficient means (no small requirement), or alternatively is willing to live with "lack" of resources, he may be able to agape-love his neighbor as he does himself. He may be able to see himself "objectively" as no more than one person among all other persons living or who will ever live. Although he will always have a more intimate sense of his own needs than of anyone else's needs, he will try to promote everyone's interests, present or future, as much as is possible. Having become good, he can rationally identify with all human interests and survive death embodied in the "ongoing rush of humanity" (*Surviving* 345). Death might even be welcomed as the final culmination of his goodness in the form of release from the "centripetal force that continued through his life to pull him back into his smaller self" (*Surviving* 15).

Johnston acknowledges the "paradoxical" nature of this immortality. Achievable by very few, out of reach of the ordinary person who given self-deception could only hope to be "good enough," in most cases, consolation must come in contemplating the example of a few saints who at least seem to have achieved it. Even more "paradoxical," he admits, is the ever-present problem of evil. Given that future humans will, like all humans on his view, be naturally disposed to favor their own interests, few are likely to be good, and very few truly good. Therefore, it is hard to see how anyone can live for the good that comes after death. At the end of a long and meticulous argument, Johnston ends on a note of despair. Given "the atrocities of the last century, atrocities that we, as a species now seem poised to outdo," given "thoroughly evil wills swept into positions of enormous power by circumstance and by their own demonic determination," given a ready supply of "collaborators seeking premium treatment for themselves in straitened and competitive circumstances," the prospect of finding salvation in future good is tenuous at best (*Surviving* 335).

Oddly enough, it is a fact of life that belies the thesis of inherent self-interest and that is of central importance in Diotima's teaching that, in the end, allows Johnston to offer some hope for the good that lives on after us. What goes unnoticed beside standard histories of war and tyranny, he points out, is "a recurrent of human history which although miraculous makes for much less memorable stories."

> I mean the collective miracle of the nurturance of the young of our species who begin life so pathetically dependent and remain that way for so long. Who can fail to be astonished at the enormous amount of distributed good will required to provide them with the material and psychological resources that gives them a chance to develop flourishing individual personalities. (*Surviving* 336)

Given an assumption of inherent self-interestedness, the natural fact of human and animal reproductive life has to seem "miraculous." With eros dismissed as of no moral value, what could make parents regularly and without thought sacrifice their own interests for the interests of their children? What would motivate citizens to pay taxes to collectively educate the young? Johnston's searcher after immortality moves from self-interest to the interests of family or clan not out of love but as an extension of his own personal identity. He is not drawn to another's beauty, or to the beautiful life they might have together, but must rely on reason to prove that they must be as valuable as he is himself. There is no talk of falling in love or mutual conversation as to how to be good. No domestic or civic initiatives, no works of art, beautiful productions, profound philosophies, or elegant theories in math and science can come from the "episodic spasm" he identifies as erotic love.

* * *

In her lessons in love, Diotima began by calling Socrates's attention to facts of reproductive and family life. He was to take note of the innate impulse in animals and humans to bond with, care for, and learn from others. He was to remember stories told of memorable acts of courage and self-sacrifice. He was to appreciate the creation of just systems of law, great works of art, systems of knowledge. The defeat of death described by Diotima requires no miraculous resuscitation of dead bodies or survival of bodiless persisting souls, but nor does it begin with the all-but-impossible task of achieving the intellectual feat of seeing all human individuals as equally valuable. It is the possibility "ever-existent" in mortal life, realized when life is lived with erotic energy. It is the striving to transcend mortality that drives sexual attraction and the bearing and rearing of human offspring as well as human cultural and social creation. It is the force of attraction that, Empedocles speculated, must have been there in rudimentary form at the very origins of the cosmos. It is neither a preordained divine plan for the workings of the physical world, nor is it a Providential premapping of the course of human history. It is participation in and affiliation with beautiful possibilities, driven by eros as the mating of lack and need with intelligence. As Diotima put it to Socrates:

> Every mortal thing is preserved in this way, not keeping itself the same forever, like gods, but by replacing what goes away or is old with something new to put in its place. In this way, Socrates, a mortal thing shares in immortality, both in body and in all other respects; it can be

> done in no other way. So do not wonder if everything naturally values its own offspring, it is for immortality that every being is moved by this eagerness and this Eros. (*Sym.* 208b1–6)

Immortality, in this sense, is not dissolving the self into nothingness, nor is it living on in one's own genetic offspring. Children must be nurtured and taught. To prosper they must live in a world in which there is beauty. As Johnston points out, extending immortality past the survival of one's own children to the survival of an extended family or tribe, or ethnic group, or even all of humanity, cannot be reassuring if humans are doomed to suffer or driven to violence and cruelty. What is reassuring is to come to see the beginnings of new life even in the midst of death and destruction. Immortality for Diotima is more than the survival of children or even of the human species, which if hopelessly compromised might evolve into new forms of life; it is the generation of good things that endure. Diotima reminds Socrates of the erotic urge of many of the men around him to "make a name for themselves and so lay down deathless fame for time eternal" (*Sym.* 209c7–9). Sacrificing oneself for a loved one, dying for one's country, braving assassination to run for political office, all are inexplicable on the basis of simple self-interest. Stronger than the will to survive or mate is the impulse to create memories of "deathless virtue" that live on as models of human excellence. Even more lasting are the creations of the poets and artists, who give life to forms of virtue and wisdom in stories and images. More lasting still are forms of justice and virtue in the management of households and the governing of cities.

Immortality in Diotima's teaching is not living on in the continuing life of a chosen people or, in less ethnocentric versions, living on in the survival of a human race doomed to endless cycles of violence and oppression. What we want when we love is for good things to come to be. None of these are products of an individual's soul or brain, but are generated in relation and intercourse with others. Sailing out on the wide sea of beauty, a person initiated into the mysteries of Diotima's "erotica" is never alone, never trapped in the narrow "arena" of his or her own "modes of presentation." In this sense, immortality is the fulfilment of the erotic urge itself as what surpasses the limitations and dangers of human life. It is not an abandonment of self and others for the nothingness of nirvana or deliverance from life seen as a cycle of suffering and pain. Instead, the immortality promised by Diotima's lessons in love is inherent in loving interactions between oneself and others in the natural world, energized by the perception of visible beauty in people and things that quickens and stimulates generation of all kinds.

From the standpoint of an individualism that assumes the essential "self-centeredness" of human beings, to care about children and future good instead of oneself has to be the anomaly of Freud's "normal exception" or, as Johnston puts it, "miraculous." The most that can be conceived is managing the impossible or all but impossible task of loving others as much as oneself. Self-centeredness, however, is not the reality of the human condition as understood by Diotima. Mortal beings are not discrete contained objects with private interior thoughts and desires, confronting other seemingly similar human objects whose private thoughts and desires are imagined only with difficulty. They are organic beings constantly changing and responding to other sentient beings and to the natural world, in dynamic relation with what is around them, painfully shrinking away from what is ugly, driven by eros to consort with and create beauty. Immortality is not based on a seemingly impossible identification with unknown members of the human species that live on after one's death in what may only be an "on-rush" of folly and cruelty. It is in the ever-present striving and overcoming of human existence, a striving that cannot be explained on the basis of physiological mechanisms of arousal and release, but only as the pull of what is beautiful and good.

Memories fade, manuscripts are lost, just states conquered, civilizations die. There is nothing in Diotima's teaching to suggest progress toward a final apotheosis, or any guarantee that human products—no matter how beautiful or good—will survive forever. There is no preset "end of man." What does endure forever, what is truly divine and the closest mortals come to immortality, is not a particular beautiful thing or a particular good thing, but the beauty that puts living things into motion. It is at this point in the lessons that Diotima warned Socrates he would have difficulty understanding.

> Into these matters of love, even you, Socrates, might be initiated, but I doubt if you could approach the final mysteries and revelations for which these are only preparation. Still, I will speak about them, she said, and will try my hardest, and you on your part must do your best to follow. (*Sym.* 209e6–210a4)

This final revelation, Diotima makes clear, does not come out of nowhere. It is not the voice of Yahweh suddenly speaking to Moses on the mountain, or the angel Gabriel reciting to Mohammad in a desert cave. It is not God choosing to reveal himself to a sinful people, or Being disclosing itself in a metaphysical haze of philosophical ecstasy. It is not a revelation you wait or hope for; it is revelation you work for. In biblical traditions, God appears first and makes clear to unenlightened and immoral humans what they must do to be saved. In Diotima's teaching, the order is reversed. The strengthening and growth of moral development comes first, preparing the way for the final revelation of divinity and the achievement of immortality. The most beautiful things—landscapes, persons, art objects, civilizations—are never perfectly beautiful or eternal, but beyond them is something else, not an object or form, but beauty itself which remains even as beautiful things come to be, die away, erode, become corrupt. Seeing this transcendent beauty and having it with you, she tells Socrates, you will not grasp onto images of virtue, and you will achieve immortality as much as any mortal can, identifying with and living in association with not an idol or image of divinity but with the timelessness of divinity itself.

* * *

The linkage between beauty and immortality is acknowledged in many religious traditions, including Judaism and Christianity. Although in Matthew's version of the parables of Jesus, the Kingdom of Heaven seems to take on otherworldly meaning, Luke and at times Mark describe the "kingdom of God" as perceivable by those who can see, and realizable in the natural world.

> Asked by the Pharisees when the kingdom of God was coming, [Jesus] answered them: "the kingdom of God is not coming with signs to be observed; nor will they say 'Look, here it is!' Or 'There!' for behold the kingdom of God is in the midst of you." (*Luke* 17:20–21)"There are some standing here who will not taste death before they see the kingdom of God." (*Luke* 9:27)

Jesus uses generative images recalling the worship of Demeter or Asherah. Heaven comes from a "sowing" of small acts. It is a mustard seed planted by a man in a garden that grows into a tree. It is the yeast that a woman "hides" in three measures of flour to leaven the whole (*Luke* 13:10–20).

The transcendence of erotic response to the beauty of a beloved, celebrated in the Bible's *Song of Songs* was revived in early Reformation texts to give sanctity to marital love. As one historian of theology put it, Protestant reformers understood eros "as the means by which

God lures God's people—exiled by sin—back to the Jerusalem of union with divine Beauty." They described "a universe caught up in deep yearning, grasping instinctively for God's loveliness, glimpsed in mirrored fragments of wonder." They envisioned the "whole world strangely linked together in a vast mystery of attraction and interdependence."[5] Although Jewish theologian Martin Buber in *I and Thou* expressly disavowed eros as "purely physical experience in which one loses the sense of the union of 'I and thou'" (87), several pages later even he claimed—not without a troubling injection of male dominance—"He who loves a woman, and brings her life to present realization in his own, is able to look in the Thou of her eyes into a beam of the eternal Thou" (106).

Mary Daly, calling in *Beyond God the Father* for the "ultimate transcendence" she saw as necessary for successful liberation movements, rejected "false gods" used to explain away evil. What was needed, she said, was the pull of a "final cause" that inspires action, not a final cause preexisting as the essence of what a woman or a man eternally and truly is, but a "catalyst that enables women and men to break out of the prison of self-destructive dichotomies perpetuated by the institutional fathers." Daly went on to find that catalyst and creative potential in the beauty of women and in love between women, inspiring new forms of spiritual expression, energizing and radicalizing feminist practice.

Catholic theologian Elizabeth Johnson describes Christian transcendence in terms similar to those used by Diotima. Divinity cannot be described or understood as an object, but is glimpsed in "fragments of beauty, goodness, and truth, both social and cosmic, facets of reality that point us in different ways to the one ineffable source and goal of all" (*She Who Is* 118). Johnson's "spirit at the center of the world" is Sophia, neglected by theologians, overshadowed by the two masculine persons of the Trinity, but ever present as an undercurrent of nontheist spirituality in Christian thought. Johnson described this "Holy Spirit" in terms reminiscent of Empedocles and Plutarch's Isis. Spirit is bringing forth and nurturing life, holding things together, and constantly renewing what the ravages of time and sin break down.

> Fertility is intimately related to her re-creative power, as is the attractiveness of sex. It is she who is ultimately playful, fascinating, pure, and wise, luring human beings into the depths of love. (*She Who Is* 135)

As Spirit Johnson's Sophia is the renewal of spring, the network of life, mutual reciprocal love. Spirit is "immanent but also transcendent," "the ultimate reality," "the highest beauty," "the source of new being in situations of death and destruction." In *Quest for the Living God*, Johnson found that divinity not in professions of faith or partaking of sacraments, but in liberation movements in Latin America, South Africa, and anywhere there is passionate resistance to oppression and a vision of more beautiful forms of life.

If one of the roles of religion is to save mortals from death, the promise of personal immortality in the form of resurrection of the body or survival of the soul offers that consolation and salvation in selfish terms. You may not be rewarded in life, but in the end your enemies will be punished. You may suffer in life, but you will be happy in heaven. You may seem to die, but you will not really die. Unlike unsaved others, you will live on in another spiritual body. The "deathlessness" promised by Diotima is not a ghostly continuation of individual mortal existence, but *theophilia*, alliance and common cause with the eternal element of transcendence in life itself. Unlike Tillich's despairing existential "courage to be" in the face of meaninglessness, it is the identification of and propagation of beauty. Unlike Buber's I/thou moment experienced as a taste of final heavenly consummation, it is participation in relations and movements that envision and create a better future. Not coming to see the self as nothing worth caring about, it is alliance with others to create new forms of life. When movements fail, when what was once beautiful decays into ugliness, it is the knowledge that the pull of beauty is an indestructible immortal element in the natural life of women and men.

Immortality here does not depend on the impossible task of giving up interest in oneself for the sake of living on in others. Nor is it a "vicarious" immortality gained through having children or living on in the memory of others? The lover of the *Phaedrus* might groom a beautiful youth to be a replica of himself. He might write plays or poetry in an attempt to win prizes or praise. The ultimate reward of friendship with divinity is more immediate and more intimate. Allied with the pull of beauty and goodness, a person is immortal, as Diotima puts it, as far as any human can be. Realization of and affiliation with this combining, creating, immortal element in natural life—"not filled-in with flesh," not as this beauty or that beauty—constitutes immortality in Diotima's final vision. In that "realization" and "affiliation," ritual and ceremonial practice have a role. Prophecy, prayer, ritual, communal celebration, and devotion all can express and further *theophilia*. But more important is love, not in its current degraded form as an "episodic spasm" of genital pleasure, not as two people mystically combining into one, not as passionless agape, but as the great daemonic spirit that is erotic desire.

* * *

It is hard not to think that new directions are needed in religious thinking. Racial and sexual divisions persist, driven from the surface of public discourse and written law, but ever more deeply rooted in structures of society and individual psychology. Militant theism and a will to suppress those who worship other gods show no sign of abating as Islamic jihad recruits fighters from all parts of the world, and evangelical Christians insist on biblically based legislation. Poverty and disease continue to fester in virulent pockets of misery that are breeding grounds for rebellion and breakdown of community ties. Wealthy commercial societies release proliferating deadening storm clouds of expensive and often useless devices and services.

Any thinking carries with it strands of older thinking; no thought stands apart from previous human experience and existing human language. At the same time, both history and language have layers: repressed understandings, insights that have been ignored, cultures whose spiritual wisdom has been lost, lines of thought neglected and misunderstood. Any thinking draws on the past, but always a choice is made as to which ideas to take up and which to allow to fall away. Essential in this process is imaginative vision that brings past moments of insight back to life.

There have been few artistic depictions of the meetings between Diotima and Socrates. David, artist of the French Revolution, made a preliminary pencil sketch for a never completed oil painting. In his version, Diotima stands awkwardly off balance, reaching out to offer Socrates a manuscript or book. Socrates, drawn with a strong well-muscled body, sits nonchalantly at ease. One hand gestures in her direction, not, it seems, to accept what she has to offer so much as to brush it away. Perhaps an older image better fits Socrates's own account of their meetings. In a small bronze relief from first-century Pompeii, he appears a less heroic figure. Short and potbellied, he stands one hand on hip, leaning on a stick. His face is clenched and his eyes are narrow in thought. Diotima, seated before him, leans earnestly forward, holding out a braided cord as if to urge him on along her line of thought. Behind them, benignly overseeing their efforts, hovers winged eros treasure casket in hand. Perhaps we might leave them there with lessons still in progress.[6]

Notes

Prologue

1. S. Weil, *The Need for Roots*, London: Ark Paperbacks, 1952, 41.

Introduction

1. See, for example, Guthrie in his 1975 *History of Greek Philosophy* (Volume IV) who argues that because, with the exception of the *Apology*, Plato could not have been a firsthand witness of any of the conversations he recreates, the dialogues should be interpreted philosophically rather than historically. Charles Kahn in "Did Plato Write Socratic Dialogue," *Classical Quarterly* 31(2), 1981, 305–320 describes the dialogues as works of dramatic fiction depicting imagined philosophical discussions so as to lay out a program for education at Plato's Academy in Athens. Daniel Anderson, in his 1993 *The Masks of Dionysus*, proposes that Diotima in the *Symposium* is a fictional character invented by Plato to explore an as-yet-unresolved contradiction between a process view of knowledge and certain knowledge of eternal Forms. More recently in their 2004 *Plato's Dialectic at Play*, Kevin and Elena Corrigan use concepts from contemporary literary theory to describe Plato as a novel-writer, "saturated with all the subtleties of his highly cultivated Athens and yet, at the same time provoking self-critical trajectories of reflective possibility outside the boundaries of his own space and time" (3).
2. One exception is Aristotle's comment that Plato wrote the *Laws* after he returned from Sicily. Plato's own *Letter VII* gives some information about his visit to Sicily but says nothing about the dates of composition of the dialogues, and its authenticity has been questioned. Otherwise, stories about Plato's life must rely on accounts written three or four centuries after his death from sources no longer extant. Catherine Zuckert's 2009 *Plato's Philosophers: The Coherence of the Dialogues* abandons any attempt at historical dating and instead reconstructs what she takes as the "fictional" dating of conversations chosen by Plato as a literary device, which he uses to tell a dramatic "story" of philosophical development.
3. See, for example, Price in *Love and Friendship in Plato and Aristotle* who finds incomprehensible Diotima's claim that contact with beauty

could make you virtuous and concludes that the philosophy of the *Symposium* is clearly deficient in comparison to that of the *Republic* or the *Phaedrus* (36–43). More recently, Gabriel Lear has a similar reaction in "Permanent Beauty and Becoming Happy," arguing that Diotima's teaching is "alien" to us and "strange" in the light of doctrines from other dialogues.

4. Nails argues that the only thing supporting the claim that Diotima is the one fictional character in the Platonic dialogues is that existing references to her all appear to be based on Plato. One other supposed bit of evidence—Diotima's reference to "other halves" refers to Aristophanes's story about hermaphrodites—is not decisive since both her remark and Aristophanes's fable could be based on preexiting myth or folklore (*Prosography* 128–129).
5. *Phaedo* 59b10. Phaedo on a visit to Phlius is asked to tell the story of what was said between Socrates and his friends in the last hours before Socrates's death. Plato, he notes, was not present, presumably because of illness.
6. Speeches and passages from writers, teachers, or orators are sometimes inserted in the dialogues. According to Diogenes Laertius, the long detailed cosmology repeated by Timaeus in the *Timaeus* is taken directly from a book by the Pythagorean philosopher Philolaus. Aspasia's funeral oration celebrating the heroism of war dead is recited verbatim by Socrates in the *Menexenus.*
7. *Metaphysics* 984–987. As Aristotle tells it, Ionian philosophers introduced the question of the origins and nature of the physical world and initiated a search for one or more primal "material causes" out of which everything is made, but identification of one basic material or materials could not explain why anything moved or came to be. Heraclitus claimed to find a motive cause in the opposition of primal elements, such as fire and earth, hot and cold, but, notes Aristotle, this cannot explain how things came to be well formed or beautifully made, which could not be simply due to chance. Hesiod gave a mythic account of Eros as a moving force in a primeval mating of earth with sky, but, argued Aristotle, love alone could account only for good, and not for what is ugly or evil. Enter Empedocles with two moving principles love and strife. But, as Aristotle notes, none of these theories solved the problem of knowledge that worried first Socrates and then Plato. How is certain knowledge possible of a physical world that is always in a state of change? Socrates turned away from natural philosophy to moral questions, looking for definitions of justice and goodness that could reliably guide political and moral beliefs. Plato at some point abandoned the search for any true knowledge of this world of sensible objects and turned to knowledge of otherworldly Ideas or Forms.
8. The plague is a matter of historical record. In 420 BC it spread north toward Greece from Ethiopia and Egypt where it killed one person in

three. As a result of Diotima's efforts, the plague did not break out in Athens until 430 BC, allowing the city to withstand the Spartan attack of 439. When the plague finally did come to Athens, it signaled the beginning of the end of the Athenian Golden Age. Pericles died of the disease, a succession of weak leaders took over, and much of the population was lost. Athens continued to fight Sparta off and on until 404, but by the time of Agathon's symposium, defeat at the hands of Sparta was imminent.

9. See Joan Connelly's *Portrait of a Priestess* (2007) for a detailed survey of archeological and inscriptional evidence documenting the many important civic and religious functions of Greek priestesses. Connelly concludes that women participated in 85 percent of religious festivities in Athens, activities that were in no way marginal but deeply embedded in civic life. She attributes the "veiling" of "women's more visible roles" in ancient Greece to the destruction of documentation during the Christian era, to bias in interpretation of the few images and inscriptions that survived the destruction of "pagan" temples in 393 AD, and, perhaps, to the erroneous "feminist principle" that ancient women were subordinate and lived sequestered in private households (*Portrait* 275–279).

1 Daemonic Eros

1. To make his distinction, Pausanias used two different mythopoetic stories of Aphrodite's origin. In Hesiod's *Theogony*, Chronos (Time) cuts off Uranus's (Heaven's) male genitals, and Aphrodite is born from Uranus's severed member which, when thrown into the sea, gives off a last bit of sperm foam. In the *Iliad*, Homer incorporates Aphrodite into the Olympian family of gods, giving her a heterosexual origin as the daughter of Zeus by an invented consort Dione. See Farnell in *Cults of the Greek States*: "Dione was a fiction that may have arisen at Dodona out of a desire to incorporate the eastern Cyprian Aphrodite in the Hellenic pantheon by adoption" (621). As for Pausanias's claim that there was vulgarity associated with Aphrodite "of the people," Farnell's conclusion after an extensive search for evidence in sanctuaries and temples of Aphrodite was that this also was invention on the part of Pausanias: "What we know is that until the declining period of Greek history the cult of Aphrodite so far as it appears in written or monumental record was as pure as that of Zeus and Athena, purer than that of Artemis, in nearly all the Greek communities." See in accord Broneer, who in the course of excavating the cave sanctuary of Aphrodite below the Acropolis sifted through extant references in ancient texts to conclude that at Aphrodite's shrines and sanctuaries cult names and titles like Ourania, Pandemos, Cypris, and Cythera were used interchangeably and in no way distinguished different deities ("Eros and Aphrodite" 52–53).

2. References to Parmenides's *Poem* are to numbered fragments of Greek text collected in *The Presocratic Philosophers*, G. S. Kirk and J. E. Raven (1957).
3. Echoes of a similar argument resurface in the *Euthyphro* where Socrates exposes the fallacy in defining piety as "pleasing the gods." Can a god be imagined as loving or wanting anything from humans? Can a god be thought to lack or need anything that would require such attention from humans? The argument is similar to Diotima's. The gods do not love or desire. Being perfectly good and beautiful, they lack nothing that humans could supply to them.
4. See Breitenberger 2007: "The Creation and Birth of Eros at the Symposium" 171–194, and Shapiro, "Eros in Love: Pederaty and Pornography in Greece" in Richlin ed. 1992: 53–72. See also Brunel, "The Scope and Temperament of Erotic Art in the Greco-roman World" in Bowie and Christenson eds. 1977. In a related change in sculpture, the old Kouros figures of young Apollos with their martial stiffness and archaic warrior stance softened to become softer, sweeter, more feminine as realistic adolescent objects of sensuous desire. Only later did the infant Cupid appear with his miniature bow, with new identities as the divine son of Aphrodite validating the secession of imperial Hellenistic and Roman rule and as a popular image of the effect of love featured in Roman bedroom art.
5. There is no material evidence of any fifth-century temples, sanctuaries, or altars dedicated to Eros as an independent deity. The only report of what might be considered official cult comes in the second century AD from the small Boetian town of Thespiae where, according to Pausanias and Plutarch, a festival named the Erodica was held with athletic games and a music competition. See Plutarch, "Dialogue of Love" in *Morales X*. See also Pausanias—"Of the gods the Thespians from the beginning honored Eros most and have a very ancient image of him in the form of a unwrought stone" (IX 27.1).
6. Foucault bases his claim that the "cultural practice" of pederasty was "accepted by opinion" and "supported in educational and military institutions" and had "religious guarantee" on, as he puts it, "the prestige of a whole literature that sang of it and a body of reflection that vouched for its excellence" (*Care of the Self* 190). There is little proof, however, that in fourth- and fifth-century Athens that "prestige" went further than conservative aristocratic circles frequented by Plato and Socrates. See Giulia Sissa's analysis of the different views of ordinary citizens of Athens in 2008 *Sex and Sensuality in the Ancient World*.
7. Epicurus's motive in claiming that he was not an atheist and that he honored the traditional state gods may have been purely expedient, to avoid a charge of impiety. In any case, the existence of gods that took no interest in human life did nothing to alter his prescription of a life focused on the judicious regulation of material pleasures and pains.

8. The root meaning of *Πόρος* (*Poros*) is what serves as a method of crossing over a boundary, such as a ford or ferry that allows messages or goods to pass back and forth over a body of water. Μητις (*metis*), usually translated as "cleverness," has a range of meanings including wisdom and strategic intelligence. *Πενία* (*Penia*) refers to need in the sense of hardship, lack of resources, and painful striving.
9. In preparation for the festival, the epistyle housing the cult statue was purified and blessed and repairs were made to its structure. With great ceremony the cult statue was washed, dressed in a purple cloak, and brought out of the sanctuary, to be paraded through the streets. See Rosenzweig 2004 for a detailed description of the archeological evidence, including remnants of the third-century stele that housed the statue and an inscription by a wealthy woman donor detailing expenditures and preparations for the event.
10. The passage has posed particular difficulties for scholars trying to make a division between Socratic and Platonic dialogues. See, for example, Price 1997: 254–255 who, in the course of defending his claim that the *Symposium* is not Socratic but Platonic, finds it necessary to explain away the "Socratic" implications of a supposed linking of all eros with "the good." Plato, he argues, is being disingenuous, concealing the fact that he has given up eudaimist psychology and presenting Socrates as "editing" his speech so as to win the prize for the best speech: "It serves Socrates's present purpose, which is to say nothing derogatory about erotic desire, that he gives no hint of any divergence or conflict of the kind that serves in the *Republic* to distinguish rational from irrational desires." Compare Rowe 2006 who argues that the dialogue is Socratic and discounts what he calls the confusing "mixing-in" of Forms at the end of Diotima's lessons.
11. The view of language here might be compared to the lengthy discussion of deficiencies in natural language and the changing meanings of words in the *Cratylus*. The subject is the perplexing problem of the knowability of what does not stay the same, whether in the natural world or in the meaning of words used to talk about the natural world. In the end, Cratylus, a disciple of Heraclitus, declares himself unconvinced that there is any eternal world of unchanging ideas to which words with fixed meanings could refer.

2 The Work of Love

1. See Sheffield 2006: 84–85 for an extensive review of what he calls "the notoriously difficult" task of explaining Diotima's turn to "birth in beauty" after the brief change of subject to the "good." A particularly convoluted example comes from C. J. Rowe 1998 "[Diotima's] whole approach is based on a deliberate and artificial conceit, i.e. one which treats human desire in general as if it possessed the

same structure as sexual desire, which she and Socrates know (and we know that they know) that it does not...We have passed from the limits of literal truth into the realm of metaphor" (251). The assumption that the passage is not to be taken literally allows Rowe to go on to take the immortality achieved in generation as no more than a stand-in for the Platonic good, eliminating any inconsistency between Diotima's teaching and Platonic doctrine (257). On Rowe's view, "the good"—both the lesser goods of children, honor, and fame and ultimately the highest "Form of the Good"—can then be substituted as the subject of the rest of Diotima's teaching, confirming the *Symposium* as precursor to the theory of Forms in the *Republic*.

2. See Pausanias's description of the temple near the Ilissos river outside of Athens, near which Socrates and Phaedrus spent an afternoon discussing love: "As for the district called the Gardens and the temple of Aphrodite there is no story told about the temple or the Aphrodite that stands near the temple, but it is square like a herm and the inscription reads: 'Heavenly Aphrodite is the first of the Fates'" (*Description of Greece* 1.19.2).
3. Ζωον, "living being" is often translated as the masculine "man." Although some of the qualities Diotima lists as never staying the same are human qualities, the sense of the passage is that change is inherent in living beings, not only in humans and certainly not always in male humans.
4. Μελέτη (studying) is sometimes translated as "cunning" other times as "recollection," neither of which seem appropriate in this context. I translate σώζει as "save" to bring out the interesting contrast with Christian "salvation" conceived as supernatural survival after death.
5. Φρόνησίν has been translated many different ways (prudence, wisdom, understanding). I have chosen mindfulness as best suited to the core meaning of thoughtful, conscious, purposeful action.
6. Most translations use "sobriety" or "temperance" for σωφροσύνη (reasonableness), indicating control of bodily appetites, and translate δικαιοσύνη (management) as "justice," suggesting an ideal heavenly icon viewed by souls liberated from the body as in the *Phaedrus*. Dikaiosine "setting things right" is from the verb δικαιόω to pass judgment. As used in Homer and other texts, Δίκη does not refer to a transcendental ideal but rather to practices of judgment, penalties, legal procedure, and recompense that communities devise to "put things right" between people in conflict. If Dike is the bringing of order or cosmos to human affairs, an individual with sophrosine brings the same order to her feelings, thoughts, and actions. See the *Charmides*, where the main subject of discussion is how to achieve that self-governance, with the Delphic "Know thyself" featured prominently. Socrates adds there the interesting suggestion

that knowledge of self may not be sufficient but that a "science of human advantage" might also be necessary.

7. See the postscript added by Dover to the 1989 edition of *Greek Homosexuality* 204–206) acknowledging the evidence that intercourse regularly occurred. Analyses of Greek pederasty from the boy's point of view are rare. A 2001 paper by Enid Bloch ("Sex between Men and Boys in Classical Greece: Was it Education for Citizenship or Child Abuse?") has had little follow-up. In Section 4, "Erotics," of *The Uses of Pleasure*, Foucault acknowledges at some length the special problems surrounding pederastic sexuality (185–215). At the same time he applauds what he takes as the "freedom" of the Greek boy's choice to participate in pederastic sex as opposed to the arranged heterosexual marriages of Greek women. He describes the boy and man as in a state of "reciprocal independence," players in an "open game." Man and boy make a "deliberate choice of one another"; the boy "consents," which Foucault defines as acquiescence that requires no violent forcing that would amount to rape. This interpretation was the basis for the 1977 petition signed by Foucault and other French intellectuals calling on the French government to eliminate all legal restrictions on child sex ("Sexual Morality and the Law" in Foucault 1988). Foucault and the other signers argued that sex with an adult was not necessarily traumatizing to a child, that a child is capable of consent, and that laws prohibiting child pornography and child sex are aimed at demonizing homosexuals, not at protecting children.
8. See *Laws* 835–841.
9. Again, Diotima's teaching causes consternation and controversy. Here the trouble is not discrepancy between "birth in beauty" and the philosopher's quest for the good, but the difference between Diotima's loving praxis and romantic and Christian ideas of "true love." See, for example, Vlastos (1973, 1991). Vlastos argued that even if there might possibly be educative value for a protégé in discourse between lovers as described by Diotima, it was completely abandoned in her description of the higher mysteries of love. Such critiques, based on a premise of individualism, take little account of the interactive and generative nature of eros as described by Diotima. On that view, lovers are not separate individuals, each with his or her separate goods, rather they create good between them. In this sense, eros is essentially different from the supposedly superior Aristotelian "philia," defined as mutually advantageous relations between friends, in which each pledges to advance the individual interests of the other. See also more recently Sheffield in *Plato's Symposium* who takes all of the examples given by Diotima as no more than the selfish use of others to gain honor for oneself and, as a result, immortality.

3 Beauty Itself

1. Translations of this important passage are often framed so as to support the interpretation of beauty itself as a Platonic form. "Not filled out with flesh" becomes "not infected" with flesh, a translation that suggests rejection of the material world. Mortal "foolishness" becomes mortal "trash," indicating a worthless and discardable body.
2. This was Herodotus's explanation, although Herodotus has a tendency to lump all foreign influences together as "Egyptian" (11.123).
3. There are similarities and dissimilarities with Aristotle's famous "unmoved mover." In the *Metaphysics* (xii, 4–10), Aristotle struggles with the problem of accounting for movement in the physical world. Given the fact that motion exists, he argues, an immaterial substance must exist that moves at the same time as it itself is unmoved, and this can only be true of an attractive object of desire. Aristotle's Mover, however, is not beauty but thought, a conclusion Aristotle reaches by logical deduction: Thought is the best form of activity; therefore the divine unmoved mover must think, but since it is "unmoved," it cannot be dependent on any external object of thought; therefore the unmoved mover must think itself, moving the heavenly spheres as an object of desire because the circularity of rational thought is the "most perfect form" of activity. "Beautiful" objects of desire then are demoted by Aristotle to mere "apparent goods" "dependent on opinion (*Meta*. xii, 6).
4. Herodotus and other Greek writers traced the coming of Aphrodite to Greece from Near Eastern origins. "I have inquired and find that the temple at Ascalon (in Syria) is the most ancient of all temples to heavenly Aphrodite. For the one in Cyprus as the Cyprians themselves admit was built in imitation of it, and that in Cythera was erected by the Phoenicians who are from this part of Syria" (1.105). Herodotus describes lines of transmission between Aphrodite and ancient Near Eastern deities acknowledging the commonality between them. "The Assyrians call Aphrodite Mylitta, the Arabians Allat, and the Persians Mitra, and also that deity is associated with the Indian deity Mitra" (I.131). Pausanias mapped out a similar trajectory, from Assyria to Paphos, and from Phoenicians in Palestine to Cythera (1.14.6).
5. Translation of this poem is from the Greek text in Page 1979, *An Introduction to the Study of Ancient Lesbian Poetry* (*Frg*. 2, 34). Translations of the other cited Sappho poems and fragments are from numbered Greek texts in Haines 1927, *Sappho: The Poems and Fragments*. Although the library at Alexandria had nine volumes of Sappho's poetry, all that survived the Christian era were bits of parchment dug from rubbish heaps, inscriptions on pieces of broken pottery, and passages from her works quoted by classical writers.

6. The title "Cypris" is often used for Aphrodite, referring to her temple on Cypris. The Greek word "κωμα," translated here as "oblivion," denotes an abnormally deep sleep induced by enchantment or some other divine influence.
7. A similar sacred sanctuary on the island of Knidos is described by Lucian of Samosata: "The floor of the court had not been doomed to sterility by a stone pavement, but on the contrary, burst with fertility as behooves Aphrodite: fruit trees with verdant foliage rose to prodigious heights, their limbs weaving a lofty vault. The myrtle, beloved by the goddess, reached up its berry-laden branches no less than the other trees that so gracefully stretched out. They never know foliage grown old; their boughs are always thick with leaves" (Pseudo-Lucian, *Erotes* 9).
8. Himmerius, quoted by Haines 152.

4 The Spirit at the Center of the World

1. References to Empedocles's poems are to the numbered fragments of Greek text collected in Kirk and Raven (1957).
2. Empedocles gives his "roots" the names of gods Zeus (fire), Hera (air), Nestis (water), Hades (earth), but his treatment indicates physical elements, not supernatural persons.
3. Events in Euripides's play take place in the seaside town of Troezen, now the modern Greek town Trinzia. A shrine to Aphrodite in the mountains gorges above the town existed in antiquity; its ruins can be found near a small Christian chapel. Little is known about Euripides's personal views on religion. His dramatic use of supernatural appearance in both *Hippolytus* and in *Medea* has indicated to some that the effect must have been meant to be satirical. I would argue that, at least in *Hippolytus*, Euripides was dramatizing conflicting views of divinity: an older divinity of fertility and generation, the new orphic and Dionysian cults, and, perhaps, a third alternative represented by Artemis, a virgin goddess allied with masculine authority.
4. In myth, Hippolytus is the offspring of a liaison between Theseus and an Amazon woman (either Antiope or Hippolyta in different versions of the story). As the story goes, Theseus either captured or seduced Hippolytus's mother, convinced her to turn traitor to her people, and then abandoned her on his return to Athens. On the occasion of Theseus's arranged marriage with the Cretan princess Phaedra, Hippolytus was shipped off to Troezen to be brought up by Theseus's grandfather. As Plutarch tells it in his biography of Theseus, the removal of Theseus and his family from Athens to Troezen was due to civil conflict over Theseus's contested parentage and his right to rule Athens. At King Aegeus's death, the sons of Pallas, Aegeus's brothers, claimed to be the rightful heirs, but

Theseus appeared claiming to be Aegeus's illegitimate son and heir. In danger from reprisals for his arrest, execution, and imprisonment of members of the Pallas family and other opponents, Theseus was advised to get out of Athens for a period of time.

5. John Godwin, Appendix A to Lucretius, *On the Nature of the Universe*, trans. R. E. Latham, New York: Penguin, 1994, 144.
6. On several occasions, disputants in Plato's dialogues use a verb form of Aphrodite, αφροδισιάζω, for sexual intercourse between men and boys. Xenophon followed, with a noun form αφροδίσια. In "Dialogue of Love," Plutarch places the term in that context. Foucault in the second volume of his *History of Sexuality*, *The Uses of Pleasure* took the term "*ta aphrodisia*" as indicating not the sexual practice of a particular group but the "classical understanding" of sex as "acts, gestures, and contacts that produce a certain kind of pleasure." Ignoring the distinction made by Plutarch between *aphrodisiaszing* and marital love, Foucault was then able to interpret the shift to idealized married love in late antiquity as no more than the old Platonic "*aphrodisiazing*" grafted on to heterosexual marriage (*Care of the Self* 199ff).
7. Plutarch does not consider the possibility of same-sex marriage, but there is nothing in his account of married love that could not apply to same-sex unions. As with Diotima, what joins married lovers together in his account is not only or even primarily the reproduction of physical offspring but rather the bond of sexual attraction and the sharing of a life in common.
8. Plutarch's negative use of "agape" can be contrasted with the later Christian use of the word to indicate a new purified Christian love free of any taint of eroticism.

5 The Highest One

1. Second Temple Jews and Christians used "syncretism" as a derogatory term for the tainting and corrupting of biblical teaching with elements taken from other traditions. For cosmopolitan Romans like Plutarch, however, it signified fruitful harmony and cooperation between different religious traditions, as indicated by the term's derivation from συγκρητισμός, or "Cretans-together," a reference to historical alliances between warring Cretan tribes who forgot their differences, found common cause, and worked together to defeat their enemies.
2. See a review of the evidence in van der Torn, "Goddesses in Early Israelite Religion" in Goodison and Morris, eds, 1998: 88–97. Inscriptions indicate that for a time the goddess Asherah was identified as a consort of Yahweh. Continued devotion to female deities is evidenced by the many sacred female figurines found in areas

inhabited by the Israelites from 1200 to as late as 550 BC. Van der Torn concludes: "It seems to me that the most attractive explanation of the majority of the Israelite figurines is to interpret them as reflections of official cult images or symbols, used outside the cult for devotional and prophylactic purposes. Replicas of cult images, as well as miniature shrines, may have kept the memory of real images and shrines alive for those who possessed them" (94).

3. Cleanthes, "Hymn to Zeus"; A. C. Pearson, *The Fragments of Zeno and Cleanthes* (*Frg.* 48), London: C. J. Clay, 1891, 274–275.
4. A strange aspect of Hosea's prophesying is the symmetry between his own family life and God's. Like Yahweh, Hosea married a prostitute, Gomer, who bore him children, left him for other men, but was eventually taken back by Hosea and forgiven.
5. Philo took Timaeus's presentation as a description of a series of actions in time by a divine Maker or Craftsman. In this he deviates from what seems to have been the understanding of Academy Platonists that Timaeus was employing a didactic device, constructing step by step a geometric figure, as might a teacher of geometry. See, for example, Aristotle's comment at *De Caelo* 279b33. "Some say that in their statements about [the world's] generation they are doing what geometers do when they construct their figures, not implying that the universe really had a beginning, but for didactic reasons facilitating understanding by exhibiting the object, like the figure, in the course of formation." Aristotle's reference in the passage is clearly to the *Timaeus*, and according to Simplicus, the "some" who are referred to as making the argument are Xenocrates and the Academy Platonists. Aristotle himself, unimpressed by what "some say," insists in his argument that the Platonic account of creation comes close to saying that the universe could somehow be both generated and eternal at the same time. Still others, such as some Gnostics, saw the similarity between the Hebrew Bible's Creator God and the "Demiourgos" of the *Timaeus* in a negative light. What was this maker of the cosmos in *Genesis* and the *Timaeus* but a blundering imitator, attempting to make out of degraded matter something that could correspond to goodness and light?
6. Philo found what he thought might be proof of the connection in the narrative frame of the *Timaeus*. In response to Socrates's request for a story about a city that actually existed rather the imaginings of "poets," Critias volunteers a "true" story of ancient Athens, told to his grandfather by Solon who heard it from priests on a visit to Egypt. These priests, suggested Philo, could have been connected with a Hebrew community in Egypt, who passed to Solon their biblical wisdom.
7. *Exodus* 31:1–6. Yahweh speaks to Moses instructing him in how to build the inner sanctum of the temple according to specifications

that reflect the wisdom and craftsmanship of God, wisdom that has in turn been implanted in "wise-hearted men."

8. "I am very dark, but very handsome, oh daughters of Jerusalem, like the tents of Kedar, like the curtains of Solomon" (*Song of Songs* 1:5). In the *Song of Songs*, the comely Southern bride takes pride in her dark beauty, but frequent Christian commentary cites the passage as indicating a woman's inferiority and her great propensity to sin, along with the magnanimity of the divine Lord who condescends to raise her up to sit by his side.

6 Demonizing the Daemonic

1. Timaeus was similarly cautious in his treatment of the traditional Greek gods. "To know or say what is the origin of the other divinities is beyond us and we must accept the tradition as of the men of old time who affirm themselves to be the children of the gods" (*Timaeus* 40d). He goes on to provide a brief summary of Hesiod's poetic genealogy of the gods in the *Theogony*, adding, perhaps in an attempt not to offend, that "since these men of old must have known what went on in their families we must believe them."
2. Although the history of early Pythagorean communities is hard to trace, there is evidence from several sources that on occasions they attempted to play a part in the rule of a city and either had to disband or flee because of public outcry or conflict with authorities. See Kirk and Raven (1957), 222–228, for a review of the sources.
3. See *Moralia* II.14, "On Superstition" (*Περὶ δεισιδαιμονίας*).
4. Aeschylus brought politics into the final scene of his *Oresteia* trilogy. At Orestes's trial for the crime of matricide, Orestes's defense council, in the person of Apollo, devises a clever argument. Orestes was not guilty of killing a parent because a mother is not a true parent but only provides a nurturing womb for the male seed. As a result, Orestes can return to Argos and take the throne as an ally of Athens, preventing an alliance of Argos with Athens's enemy Sparta. The Eumenides protest the wrongness of this award of a throne "dripping with blood," but are pacified by Athena who offers them a substitute position as domestic deities of the hearth.
5. Philo takes liberties here. The "ha-Satan" who comes to tempt Job is an agent of the Lord, as is the "accuser" of the high priest Joshua in Zechariah's apocalyptic vision (*Zech.* 3:1–2). On other occasions, however, "satan" without an article means simply "adversary" as used generally for persons and peoples who are "enemies" or "opposers" (*I Sam.* 29.4; *II Sam.* 19:22; *I Kings* 5:4, 11:14, 23, 25). The *Septuagint's* translation of "ha-Satan" as διάβολος from the Greek verb διαβάλλο—to slander or falsely accuse—introduced a note of evil intention. What in the Hebrew Bible's *Job* was a divine test or

accusation sent by God is personified as an independent adversary of God, an agent gone wrong and a rebel against God's rule. The *Septuagint's* "Satan" as diabolos or "slanderer" becomes in English "the Devil," an individual existent and powerful demonic being.

6. See *Book of Jubilees*, *Apocalypse of Adam*, *Apocalypse of Moses.*
7. Communal sacred "love" meals continue to be held in Greek towns. An animal, donated by a citizen, is blessed by the priest, paraded through the street, and killed observing various ritual procedures. The meat is cooked in communal vats and shared by the townspeople at a communal feast. In Paul's day, some of the meat was given to local butchers to sell, with the result that in many instances little meat was available that had not been previously blessed by a priest or priestess, posing a dilemma for demon-fearing Christians.
8. *Histories* (1.199). Herodotus describes an account he heard of a "foul" custom at a temple in Babylonian to Aphrodite under the Assyrian name Mylitta. Unmarried women of respectable families were sent to line up outside the temple for an initiatory night of love. Obligated to go with the first man who threw money into their laps, they had sex with him and then returned home. Women who were fair and tall were chosen immediately, said Herodotus; less attractive girls might wait much longer remaining lined up for as long as three or four years. Something like this practice, he added, he heard went on in parts of Cyprus. Herodotus's unreliability when describing exotic locales by hearsay as well as details that strain credulity did nothing to prevent the story being repeatedly cited, both by early Christians and by later scholars and historians, as proof positive that sexual acts were involved in the worship of pagan goddesses.
9. See Stephanie Budin's *The Myth of Sacred Prostitution*, a painstaking and through review of the lack of any reliable non-Jewish or Christian historical or archeological evidence for anything resembling ritual sex or sacred prostitution in the temples of either the Near East or Greece and Rome, along with an account of the chain of mistakes and inadvertences that led to this misconception.
10. By all accounts, Origen's own asceticism was extreme. Although there have been questions whether the story was a calumny made up by theological rivals, he was said to have submitted to self-castration in order to free himself from sexual desire.
11. See Brakke 2006 for an extensive review of the various practices.
12. See *CG* 8.14. Apuleius elaborated on stories of invisible demonic creatures—some of whom once inhabited bodies and some like Eros who never had bodies—beings who inhabit the air, feel emotions including anger and hate, and demand to be placated in violent rites and rituals. Augustine was exposed to Apuleius's demonologies as early as age 15 when he spent a year studying at the University town of Madaurus, Apuleius's birthplace where his writings were well known.

13. Cybele, an Anatolian mountain deity, with aspects similar to the Minoan "Mistress of Animals," as well as to the Greek Demeter and Rhea, was introduced in Greece as early as the sixth century BC. In the third century AD, her cult was popular in Rome. Augustine was particularly horrified by her transgendered priests (*CG* 7.26). He was more than ready to believe the rumors that castration rituals went on in her temples, even though he admitted that "even under torture," no participant had ever confessed to them. Augustine also had harsh things to say about the African Caelestis or "Heavenly One." In Milan, he seconded the reprimand given to his mother Monica for bringing "impure" offerings of cakes and bread reminiscent of offerings to Caelestis into Ambrose's church (*Confessions* 6.2). Venus, the Roman Aphrodite, was particularly vile in his eyes and he regularly refers to her by derogatory epithets: Cloacina, Volupta, Libertina (*CG* 4.8).
14. Augustine's reference is to Socrates's fanciful etymology of "daemon" in the *Cratylus* (398b). In the course of a long discourse on the origins of names, Socrates decides that "daemon" is derived from "knowing" or "wise," a conclusion with little to recommend it linguistically.

7 Saint Augustine and Concupiscence of the Flesh

1. There is no extant copy of Cicero's *Hortensius*, the book that provided Augustine with a popularized version of Greek philosophy. As an African colonial subject struggling to get the education that would advance him in imperial circles, Augustine did not know Greek and had to read what was available on or by Plato either in translation or in secondary sources.
2. In Rome concubinage was a recognized legal arrangement in which a man cohabitated with a woman with the understanding that any children born would not be legal heirs or inherit property.
3. *Retractions*, Book II, Chapter 53.
4. Paula, Eustochium's mother, was one of several aristocratic Roman women who clustered around Jerome in Rome. When rumors of improper relations spread about Jerome's relations with some of his women followers, Jerome was forced to leave Rome. Paula and her daughter went with him, providing for his every need, buying his books, helping him translate the Hebrew Bible, editing and copying his writings. An additional cause for public opinion to turn against Jerome in Rome had been the death of Eustochium's sister Blaesilla who Jerome persuaded to adopt the same punishing ascetic lifestyle he pressed on Eustochium. When the girl died after four months of his prescribed regime, Jerome insisted that her mother Paula not grieve, and public opinion in Rome had been further outraged.

8 The Eclipse of Beauty

1. Nebel's title, "*The Event of the Beautiful*" referred to his critique of divinity as static Being. Instead, he argued, God as Beauty had to be an active dynamic happening.
2. In *Seeing the Form*, Balthasar mingles his voice with Nebel's, incorporating quotes from *The Event of the Beautiful* into his own line of thought as he works through and corrects for Catholicism what he calls the "best a Protestant could do as a theological aesthetics." At first he agrees with Nebel, then begins to deviate, cordoning off the Incarnation from any pre-Christian intimations of mythical or tragic beauty.

9 Religion without God

1. Υποδοχήν (bearer) is often translated as "Receptacle" and τιθηνην (caregiver) as "Nurse," but both words have somewhat wider connotations. Υποδοχήν means, in addition to being a receptacle, to become pregnant and to undertake or promise; τιθήνην more generally means to take care of or put in place.
2. See Derrida and Eisenman 1997 for Derrida's proposal for a Paris garden's harp-like structure as metaphor for the *Timaeus*'s khora; see also Derrida 1995a "Khora.
3. See Derrida 1995b, *The Gift of Death.* See also scattered references to God or religion in many of Derrida's writings, including "Faith and Knowledge" and "Spectors of Marx," in David Shakespeare's 2009 *Derrida and Theology.* See also John Caputo's 1997 *The Prayers and Tears of Jacques Derrida* for a lucid treatment of Derrida's intricate endlessly nuanced and mutating interactions with religious texts and theological concepts.
4. See the writings collected in *A Brief Inquiry into Sin and Faith.*
5. See *Varieties of Religious Experience* 498–499. James's personal belief, which he expresses at the very end of *Varieties*, is reminiscent of Diotima's daemonic pathways. Other worlds of consciousness exist and have meaning for our lives. For the most part, these two worlds remain separate, although they are continuous at certain points.
6. Both Johnson and Rahner struggle as Catholic theologians with the weight of mainstream Christian belief focused on biblical history and the figure of Jesus as the unique and one-time incarnation of divinity. At times Johnson locates "Spirit" within Christian theology as the Holy Spirit, the "third" of the Christian Trinity along with God the Father and his deified son Jesus. At the end of *Quest for the Living God* in answer to the theist question—"Does this mean there is no predetermined providence or plan for the universe?"—she returns to the biblical image of God as Lover but in a somewhat different mode. As Providence, she said. God "grants" the gift of freedom that allows

the universe to decide for itself what it will be. He does not so much have a plan as a "vision" of a community of love that he reveals in Jesus. This refixing of Christian belief onto the unique Christian event of the advent of Jesus was not enough to save her from condemnation by the Catholic Church and a ban on the teaching of *Quest for the Living God* in Catholic seminaries and schools.

7. In *The Idea of the Holy*, Lutheran historian of religion, Rudolf Otto, found the early "idea" of the holy in the sense had by "primitive" people of something "uncanny" or numinous attached to particular times or places. But, he argued, only when that sense coalesces into an entity, a figure, a god does true religion begin to emerge, and only when that figure fuses with moral authority as a guarantee of oaths or defender of the sanctity of marriage does religion reach maturity. In the *Problem of Pain*, Christian theologian C. S. Lewis describes that advance coming as a miraculous surprise with the Jews. The Hebrew Bible represents a "moralistic theism" without which religion involves "obscenities and barbarities" (12). Those who refuse to associate morality with the sense of the "holy" in this sense remain "savages," worshipping sexuality, or the dead, or the life force, or the future (15).
8. See Dean Hammer, *The God Gene.* Hammer argues that "feeling close to a powerful spiritual source" can be shown to be "a complex amalgamation in which certain genetically hardwired biological patterns of response and states of consciousness are interwoven with social, cultural, and historical threads" (5). As such "spiritually" or "self-transcendence" can be measured and correlated with a gene that involves a protein that affects the brain in similar ways as certain psychoactive drugs. On this view, religious belief mimics drug addiction as the numbing firing of electrochemical currents through nerve cells.
9. See especially chapter 5.
10. For a review of recent work, see Steven Johnson 2001, and also Scarfe and Christian eds. 2013.
11. See, for example, Robert Chapais, *Primeval Kinship.* Chapais argues that pair bonding is the determinative moment that leads to hominid development, allowing for the better care and brain development of young, affiliation between extended family members, and alliances between groups that reduce the destructive tyranny of alpha males.
12. Turner argues that simple genetic replication—he uses the metaphor of tinkering until an arrangement results that works—is insufficient to explain organic functioning. The "tinkering" of random combinations of genes requires an "apprentice," an element of natural purpose and interior and exterior design that can be seen in nonhuman animals like bees, termites, and beavers, as well as in humans.

13. See *Process and Reality*, Chapter II, especially p. 344, also *Science and the Modern World*, where Whitehead argues that the object of religion is not primarily the enforcement of morals, but the sense of something beyond ordinary reality that gives meaning and that is the "ultimate quest" as something like "the harmonious adjustment of complex detail" (192). There are differences. Whitehead continued to take most of his repertoire of concepts and language for "God" from Christianity. God is often "He" and described in anthropomorphic terms as having "vision," "knowledge," "desires," and "purposes." In addition, Whitehead takes religious experience as predominantly solitary, an individual experience of "subjective intensity" or "aesthetic" appreciation of the momentary "harmonization of all inputs" (*Process and Reality* 27; *Adventures in Ideas* 252).

10 Social Virtue

1. Here Paul uses the Greek word used by Plato in the *Lysis* for the slave guardians who come to the gymnasia to see young boys safely home.
2. For the text see www.zenit.org/article-16955. As many critics have pointed out, the comparison with Islam is hardly based on historical fact, given that as much, and perhaps more, violence accompanied Christian conquests as it did the spread of Islam.
3. Aristotle's defense of slavery—some peoples are "natural" slaves lacking full rational capacity, therefore should be under the rule of rational masters—was used to defend slavery well into the nineteenth century. Aristotle himself can be given credit for realizing that distinguishing irrational natural slaves from rational masters presented something of a problem. Simply being conquered in war could not be the criterion, since Greeks did not always win their battles. Nor, given the ethnically diverse Hellenistic culture evolving under Macedonia rule, was there any mark like skin color that could serve as an indication of irrationality. In the end, Aristotle settled somewhat unhappily on Greek-speaking as the test of rationality.
4. Karen Armstrong in her influential *A History of God* finds that Hebrew prophets' calls for social justice in times of crisis amounted to the innovation of "an egalitarian and socialist ethic" and that "the Jews would be the first to establish a welfare system" (48). A recent version of the argument is given by Nicolas Wolterstorff 2010. God and justice, argues Wolterstorff, are inextricably "intertwined" in calls by Hebrew prophets to care for the vulnerable, in the Kingdom of God as proclaimed by Jesus, and in Augustine's "caritas." The concept of a loving Father in Heaven is essential, argues Wolterstorff, if one is to support the idea of equal human worth that must underlie any theory of rights.

5. In *Leviticus*, for example, "You shall not hate your brother in your heart, but you shall reason with your neighbor, lest you bear sin because of him. You shall not take vengeance or bear any grudge against the sons of your own people, but you shall love your neighbor as yourself" (19:17–18). The passage is repeated by Jesus (*Matthew* 22:37–41). The gospel of *Luke* (10:29–37) adds a call for Christian outreach to the Samaritans, a regional version of Judaism condemned as heretical by the Jerusalem priesthood. When Luke's Pharisee goes on to ask Jesus, "And who is my neighbor," Jesus answers with a parable in which a Samaritan shows more compassion for a wounded man lying at the side of the road than either a Levite or a priest of the Jerusalem temple. The comparison is pejorative: even a mixed-race Samaritan is more kind to a Jew than a fellow Jew.
6. See Mark Johnston's contemporary version of the argument in *Saving God*. We are "fallen creatures as evidenced in the story of Eden, locked into the closed world of selfish desires." For Johnston, "altruistic" agape love is the mark of a "higher" ethical life in which you treat others with reciprocity, an idea Johnston says is "madness" if detached from religious tradition (90).
7. See, for example, John Rawls's "difference principle" in *A Theory of Justice* and his subsequent struggle to define the "veil of ignorance" that is supposedly necessary if one is to make impartial judgments about fair and just social and political arrangements. The assumption is that in a natural state of involvement in human affairs, human selfishness always skews decisions in one's own favor, and it is only by imagining a fictional and purely imaginary stance—that you do not know who you are—that a person can define the criteria for just political arrangements.
8. This very simply stated is the solution of Ronald Dworkin as worked out in detail in *Justice for Hedgehogs*. Marking the line between self-interest and the interests of others, he distinguished between the "freedom" to do anything one wants, which can be curtailed in the interest of others, and individual "liberties" such as freedom of speech, conscience, and due process, which cannot be curtailed in the interest of others and that must be protected by judicial review. Although in pursuit of an illusive formula of adjustment between equality and freedom, Dworkin proposes a "partnership" model of democracy rather than a "majoritarian" model, partnership as he describes it is not based on affiliations or relations between citizens but rather on arms-length recognition of what every individual citizen's "dignity" requires, namely the standard individual liberties of speech, opinion, conscience, acquiring property, and the vote (382ff).
9. There have been a number of variations of a "communitarian" opposition to the liberal social views of Rawls, Dworkin, and others,

in which instead of individuals being free to choose identities and life-styles, identity is seen as fixed within social structures, forms of life, or conceptual universes beyond individual choosing (see Sandel 1982, *Liberalism and the Limits of Justice*, or Bellah 1985, *Habits of the Heart*). Although relation is seen here as prior to self, social structures that pattern relations are seen as existing apart, not only from willful "choice" but also from the motivations and desires of the individuals who enter into relations.

11 The Problem of Evil

1. Until the late nineteenth century, "to sex," derived from the Latin verb "secare" "to cut," was to divide organisms into male and female with a primary application in animal husbandry. It was "sex" in the new twentieth-century Freudian sense—activities that result in a particular kind of genital pleasure—that Michel Foucault took as his subject in *History of Sexuality* and projected back on classical, medieval, and early modern experience.
2. Compare Foucault who told the story of "repression," not as social necessity but as the assertion of punitive institutional power over individual subjects. As a result, he advocated not the restriction and efficient management of sexual pleasure but the freedom of individual subjects to engage in any kind of "sexual" activity that gives them pleasure, subject only to individual choice and styling in "uses of pleasures" and "care for the self."
3. See Freud's vehement rejection of the Christian "Love your neighbor as yourself"(*Civilization* 56). Why should we love a neighbor as much as we love ourselves? How can it be right to love someone without any merit, love him even if he is bad? How can it be fair to put a stranger on a par with family and friends? Any such love, said Freud, dispensed to all humans, is so diluted that it is no love at all, reducing the Christian teaching on love to absurdity.
4. This is the point on which Herbert Marcuse in *Eros and Civilization* takes issue with Freud. Freud, argued Marcuse, missed the possibility of "nonrepressive sublimation" of eros. Although Freud acknowledged the possibility of "aim-inhibited sexual impulses" directed at parents, children, friends, and affection in marriage, he did not see these as a possible channel for the extension of erotic energy in wider social relations. Marcuse's extended social eros, however, continues to be driven by Freud's pleasure principle. His vision of liberation is a "libidinal" society in which work is reduced to a minimum, leaving the majority of time open for pleasure (189). Although Marcuse agrees that instincts can "develop" without repression, and that there might be a "self-sublimation of sexuality," that sublimation is restricted to the availability of a wider field

of individual pleasures. Without the response to beauty, which for Diotima is at core of erotic response, this is the only development possible. The creation of beautiful practices and forms of life as well as the idea that demanding productive work might be erotically driven is ruled out.

5. One approach to these passages has been to attribute them to Freud's own unresolved complexes relating to his mother. See, for example, J. Raphael-Lieff, "Freud's Prehistoric Matrix—Owing 'Nature' a Death," *International Journal of Psychoanalysis*, 88(6), 2007, 1345–1373. I would argue that deeper intellectual and cultural anomalies continue to trouble Freud's attempts to successfully analyze women's experiences in libidinal terms. From his youth, Freud had a fixation on antiquities, surrounding himself with emblems and images from ancient Egypt, Greece, and Assyrian, as if their mysterious presences communicated intimations of another world of meaning and being hidden behind classical and biblical culture.
6. In *Upheavals of Thought*, Nussbaum identifies corrective "remedies" for eros in Augustine, Dante, Spinoza, some "faith-based" writings, as well as in Diotima's teaching in the *Symposium*, but her interpretation of Diotima's discourse is Platonic as it was in the *Fragility of Goodness.* She notes in *Upheavals* that she has somewhat revised her account of Diotima's discourse in *Fragility of Goodness* to give more weight to creativity and has changed her view that Diotima thinks that all beautiful things are the same. She also somewhat nuances her claim in *Fragility of Goodness* that a lover who "ascends" with Diotima would come to see "earthly need and longing" as "mortal rubbish," or that the Diotimean lover would feel disgust at the sight of the disadvantaged, considering them just so much "mud, hair, and dirt." She admits that the word "phluaria" at *Symposium* 211e means "trivia" not rubbish and does not indicate disgust (497, n19). But she continues to appeal to Plato's repudiation of pity in the *Republic* to support her original assertion that the Diotimean lover must look down on the hungry and sick as fools. Assuming that Diotimean initiation is ascent to an intellectual goal as in the *Phaedrus*, she continues to interpret what the Platonic lover loves in the loved one as what is good for the lover. See my critique in Nye (1990).
7. See Beldon Lane, *Ravished by Beauty*, for the argument that spirituality inspired by beauty in some strains of Christian thought supports efforts to save the environment. Lane warns, however, against a sacred eros severed from theistic transcendence and power. He reiterates that opening the door to sacred eros can be dangerous because it might lead to "idolatry" and to even more rigid control of sexual practice. As Lane puts it, although "a spirituality of desire completes and enriches a spirituality of awe," "this passionate longing for God's Beauty possesses vast resources for deluding itself" (28, 40).

12 The Defeat of Death

1. The reference is probably to the ancient Mesopotamian practice of exposing the bones of the dead to be picked clean by vultures.
2. See, for example, *Isaiah* 27:12–13. “In that day from the river Euphrates to the river of Egypt, the Lord will thresh out the grain and you will be fathered one by one, Oh people of Israel. And on that day a great trumpet will be blown, and those who were lost in the lands of Assyria and those who were driven out of the land of Egypt will come and worship their Lord on the holy mountain of Jerusalem.”
3. In an attempt to make sense of resurrection, Maimonides offered an elaborate scenario. A first resurrection was bodily. This allowed judgment to be passed on identifiable individuals. After that judgment, a second death would occur resulting in either eternal life for the righteous as an embodied soul or the final death of the soul in case of negative judgment.
4. Johnston names as “major religions” Christianity, Judaism, some varieties of Buddhism, and Hindu Vedanta. Islam is not mentioned, although it would have to be considered major in terms of membership and international presence.
5. Lane, *Ravished by Beauty*, 216.
6. David’s sketch is in the National Gallery of Art in Washington, DC. The bronze relief is in Naples, at the Museo Nazionale.

Bibliography

Alcoff, Linda and Caputo, John D., eds. *Feminism, Sexuality, and the Return of Religion*. Indianapolis: Indiana University Press, 2011.

Anderson, Daniel. *The Masks of Dionysus*. New York: SUNY Press, 1993.

Aquinas, Thomas. *Summa Theologica*. New York: Oxford University Press, 2003.

Armstrong, Karen. *A History of God: The 4000-Year Quest of Judaism, Christianity, and Islam*. New York: Balantine Books, 1994.

Augustine. *Basic Writings of Saint Augustine*, ed. Whitney Gates. New York: Random House, 1948.

———. "De Nuptiis et Concupiscence." *Retractions*, Book II, Chapter 53. www.newadvent.org/fathers/15072

Balthasar, Hans Urs von. *The Glory of the Lord, Volume 1: Seeing the Form*. San Francisco: Ignatius Press, 1982.

———. *Engagement with God*. San Francisco: Ignatius Press, 2008.

Bellah, Robert. *Habits of the Heart: Individualism and Commitment in American Life*. Berkley: University of California Press, 1985.

Berve, Helmut and Gruden, Gottfried. *Greek Temples, Theaters and Shrines*. New York: Harry Abrams, 1963.

Bloch, Enid. "Sex between Men and Boys in Classical Athens: Was It Education for Citizenship or Child Abuse?" *Journal of Men's Studies* 9, 2 (Winter 2001).

Boman, Thorleif. *Hebrew Thought Compared to Greek*. London: Westminster Press, 1960.

Botkin, Daniel. *Discordant Harmonies*. Oxford: Oxford University Press, 1990.

Bottero, Jean. *Religion in Ancient Mesopotamia*, trans. Teresa Lavender Fagio. Chicago: University of Chicago Press, 2001.

Bowie, Theodore R. *Studies in Erotic Art*. New York: Basic Books, 1970.

Brakke, David. *Demons and the Making of the Monk: Spiritual Combat in Early Christianity*. Cambridge, MA: Harvard University Press, 2006.

Breitenberger, Barbara. *Aphrodite and Eros: The Development of Erotic Mythology in Early Greek Poetry and Cult*. New York: Routledge, 2007.

Broneer, Oscar. "Eros and Aphrodite on the North Slope of the Acropolis in Athens," *Hesperia* I (1932): 31–55.

Buber, Martin. *I and Thou*. New York: Charles Scribner's Sons, 1937.

Budin, Stephanie. *The Myth of Sacred Prostitution*. Cambridge: Cambridge University Press, 2008.

Burkert, Walter. *Greek Religion*, trans. John Raffan. Cambridge, MA: Harvard University Press, 1985.

Caputo, John. *The Prayers and Tears of Jacques Derrida: Religion without Religion*. Bloomington: Indiana University Press, 1997.

———. *On Religion*. London: Routledge, 2001.

———. *The Weakness of God: A Theology of the Event*. Bloomington: Indiana University Press, 2006.

———. *Prayers and Tears of Jacques Derrida*. Bloomington: Indiana University Press, 1997.

Chapais, Robert. *Primeval Kinship*. Cambridge, MA: Harvard University Press, 2008.

Cleary, John and Gurtler, Gary, eds. *Proceedings of the Boston Area Colloquium in Ancient Philosophy*. Leiden: Koninklijke Brill Nv, 1998.

Connelly, Joan Breton. *Portrait of a Priestess: Women and Ritual in Ancient Greece*. Princeton, NJ: Princeton University Press, 2007.

Corrigan, Kevin and Glazov-Corrigan, Elena. *Plato's Dialectic at Play*. University Park: Penn State University Press, 2004.

Cross, Frank Moore. *Canaanite Myth and Hebrew Epic: Essays in the History of the Religion of Israel*. Cambridge, MA: Harvard University Press, 1973.

Daly, Mary. *Beyond God the Father*. Boston, MA: Beacon Press, 1985.

Davidson, James. *The Greeks and Greek Love*. New York: Random House, 2007.

Dawkins, Richard. *The Blind Watchmaker*. New York: Norton, 1986.

———. *The God Delusion*. Boston, MA: Houghton Mifflin, 2008.

Deming, Will. *Paul on Marriage and Celibacy: The Hellenistic Background of I Corinthians 7*. Grand Rapids: William Erdmans, 2004.

Dennett, Daniel. *Darwin's Dangerous Idea: Evolution and the Meanings of Life*. New York: Simon & Schuster, 1995.

———. *Breaking the Spell: Religion as a Natural Phenomenon*. New York: Viking, 2006.

Derrida, Jacques. "Khora" in T. Dutoit, ed. *On the Name*. Stanford: Stanford University Press, 1995a.

———. *The Gift of Death*, trans. David Wills. Chicago: Chicago University Press, 1995b.

Derrida, Jacques and Peter Eisenman. *Chora L Works*. New York: Monacelli Press, 1997.

Dillon, John. *The Middle Platonists*. Ithaca, NY: Cornell University Press, 1977.

Dillon, Matthew. *Girls and Women in Classical Greek Religion*. New York and London: Routledge, 2002.

Dover, Kenneth J. *Greek Homosexuality*. Cambridge, MA: Harvard University Press, 1978.

Du Bois, Page. "The Platonic Appropriation of Reproduction" in Nancy Tuana, ed. *Feminist Interpretations of Plato*, 139–156. University Park: Penn State University Press, 1994.

Dworkin, Ronald. *Justice for Hedgehogs*. Cambridge, MA: Harvard University Press, 2011.

———. *Religion without God*. Cambridge, MA: Harvard University Press, 2013.

Eco, Umberto. *The Aesthetics of Thomas Aquinas*, trans. Hugh Bredin. Cambridge, MA: Harvard University Press, 1985.

Farnell, Lewis Richard. *The Cults of the Greek States*, vol. 2. Cambridge, UK: Clarendon Press, 1896.

Flacelière, Robert. *La Vie quotidienne en Grèce au siècle de Péicles*. Paris: Hachette, 1959.

Foucault, Michel. *The History of Sexuality*, trans. Robert Hurley. *The Use of Pleasure*, vol. 2. New York: Pantheon, 1985.

———. *The Care of the Self*, vol. 3. New York: Vintage Books, 1986.

———. *Politics, Philosophy, Culture: Interviews and Other Writings*, trans. Allan Sheridan. New York: Routledge, 1988.

Fox, Robin Lane. *Pagans and Christians*. New York: Alfred Knopf, 1987.

Freeman, Charles. *A New History of Early Christianity*. New Haven, CT: Yale University Press, 2009.

Freud, Sigmund. *Future of an Illusion*, trans. James Strachey. New York: Norton, 1961a.

———. *Civilization and Its Discontents*, trans. James Strachey. New York: Norton, 1961b.

———"Female Sexuality," 1931 in *Standard Edition of the Complete Works of Sigmund Freud*, vol. XXII, ed. and trans. James Strachey. London: Hogarth Press, 1968.

Friedrich, Paul. *The Meaning of Aphrodite*. Chicago: University of Chicago Press, 1978.

Gaca, Kathy. *The Making of Fornication: Eros, Ethics, and Political Reform in Greek Philosophy and Early Christianity*. Berkeley: University of California Press, 2003.

Gere, Cynthia. *Knossos and the Prophets of Modernism*. Chicago: University of Chicago Press, 2009.

Goodison, Lucy and Morris, Christine, eds. *Ancient Goddesses: The Myths and the Evidence*. Madison: University of Wisconsin Press, 1999.

Guthrie, William Keith Chambers. *History of Greek Philosophy*, vol. IV. Cambridge: Cambridge University Press, 1975.

Haines, C. R. *Sappho: The Poems and Fragments*. London: Routledge, 1927.

Halperin, D. M., Winkler, J. J., and Zeitlin, F. I., eds. *Before Sexuality: The Construction of Erotic Experience in the Ancient Greek World*. Princeton, NJ: Princeton University Press, 1990a.

———. "Why Is Diotima a Woman?" in *100 Years of Homosexuality: and Other Essays on Greek Love*. London: Routledge, 1990b.

Hammer, Dean. *The God Gene: How Faith Is Hard-Wired into Our Genes*. New York: Doubleday, 2004.

Harris, Sam. *The Moral Landscape: How Science Can Determine Human Values*. New York: Free Press, 2010.

Heid, Stefan. *Celibacy in the Early Church: The Beginnings of a Discipline of Obligatory Continence for Clerics in East and West*. San Francisco: Ignatius Press, 2000.

Heidegger, Martin. *An Introduction to Metaphysics*. New Haven, CT: Yale University Press, 1959.

Hubbard, Thomas, ed. *Homosexuality in Greece and Rome*. Berkley: University of California Press, 2003.

Irigaray, Luce. *The Ethics of Sexual Difference*, trans. Catherine Martin. Ithaca, NY: Cornell University Press, 1986.

———. "Sorcerer Love: Diotima's Speech" trans. Kuykendall in *Feminist Interpretations of Plato*, ed. N. Tuana. University Park: Pennsylvania State University Press, 1994.

James, William. *Varieties of Religious Experience*. New York: Random House, 1929.

Johnson, Elizabeth A. *She Who Is: The Mystery of God in Feminist Theological Discourse*. Hertford: Crossroad Press, 1993.

———. *Quest for the Living God*. New York: Continuum, 2007.

Johnson, Luke Timothy. *Among the Gentiles: Greco-Roman Religion and Christianity*. New Haven, CT: Yale University Press, 2009.

Johnson, Steven. *Emergence: The Connected Lives of Ants, Brains, Cities, and Software*. New York: Scribner, 2001.

Johnston, Mark. *Surviving Death*. Princeton, NJ: Princeton University Press, 2010.

———. *Saving God: Religion after Idolatry*. Princeton, NJ: Princeton University Press, 2011.

Kahn, Charles. *Plato and the Socratic: The Philosophical Use of a Literary Form*. Cambridge: Cambridge University Press, 1998.

Kaufmann, Stuart. *Reinventing the Sacred: A New View of Science, Reason, and Religion*. Philadelphia: Basic Books, 2008.

Kirk, G. S., Raven, J. E., and Schofield, M. *The Presocratic Philosophers*. Cambridge: Cambridge University Press, 1957.

Kondoleon, Christine, ed. *Aphrodite and the Gods of Love*. Boston, MA: Museum of Fine Arts, 2011.

Kraemer, Ross Shepard. *Her Share of the Blessings: Women's Religion in the Greco-Roman World*. Oxford: Oxford University Press, 1992.

———. *Women's Rituals in the Greco-Roman World*. Oxford: Oxford University Press, 2004.

Kristeva, Julia. *Revolution in Poetic Language*. New York: Columbia University Press, 1984.

Lane, Belden C. *Ravished by Beauty: The Surprising Legacy of Reformed Spirituality*. Oxford: Oxford University Press, 2011.

Lear, Gabriel. "Permanent Beauty and Becoming Happy" in J. H. Lesher, Debra Nails, and C. C. Sheffield, eds. *Plato's Symposium: Issues in*

Interpretation and Reception. Cambridge, MA: Center of Hellenic Studies, Harvard University, 2006.

Lewis, C. S. *The Problem of Pain*. New York: HarperCollins, 1940.

Lewis, Richard Farnell. *The Cults of the Greek States*. Oxford: Clarendon Press, 1896.

Marcuse, Herbert. *Eros and Civilization: A Philosophical Inquiry into Freud*. Boston, MA: Beacon Press, 1955.

Malraux, Andre. *The Metamorphoses of the Gods*. Garden City, NY: Doubleday, 1960.

Marion, Jean-Luc. *God without Being: HorsText*, trans. Thomas Carlson. Chicago: University of Chicago Press, 1991.

———. *The Erotic Phenomenon*, trans. Stephen Lewis. Chicago: University of Chicago Press, 2007.

Miller, Patrick. *Religion of Ancient Israel*. Westminster: John Knox Press, 2000.

Morgan, Michael. *Platonic Piety: Philosophy and Ritual in 4th Century Athens*. New Haven, CT: Yale University Press, 1990.

Nails, Debra. *The People of Plato: A Prosography of Plato and Other Socratics*. Indianapolis: Hackett Publishing, 2002.

Nussbaum, Martha. *The Fragility of Goodness: Luck and Ethics in Greek Tragedy and Philosophy*. Cambridge: Cambridge University Press, 1986.

———. *Upheavals of Thought: The Intelligence of Emotions*. Cambridge: Cambridge University Press, 2001.

Nye, Andrea. "Rethinking Male and Female: The Pre-Hellenic Philosophy of Mortal Opinion." *Journal of the History of European Ideas* 9, 3 (1988): 261–280.

———. "The Hidden Host: Irigaray and Diotima at Plato's Symposium." *Hypatia* 3, 3 (Winter 1989). This article is reprinted in Nancy Fraser and Sandra Bartky, eds. *Revaluing French Feminism*. Bloomington: Indiana University Press, 1992, and also in Nancy Tuana, ed. *Feminist Interpretations of Plato*. University Park: Pennsylvania State University Press, 1994.

———. *Words of Power: A Feminist Reading of the History of Logic*. New York: Routledge, Chapman Hall, 1990a.

———. "The Subject of Love: Diotima and Her Critics." *Journal of Value Inquiry* 24 (1990b): 135–153.

———. "Philosophy: A Woman's Thought or a Man's Discipline? The Letters of Heloise and Abelard" *Hypatia* 7, 3 (Summer 1992): 1–22. Reprinted in *Hypatia's Daughters: 1500 Years of Women Philosophers*, ed. Linda Lopez McAlister. Bloomington: Indiana University Press, 1996.

———. "Irigaray and Diotima at Plato's Symposium." *Feminist Interpretations of Plato*, ed. Nancy Tuana. University Park: Penn State University Press, 1994.

Oakesmith, John. *The Religion of Plutarch: A Pagan Creed of Apostolic Times*. London: Longmans, 1902.

Otto, Rudolf. *The Idea of the Holy: An Inquiry into the Non-rational Factor in the Idea of the Divine and Its Relation to the Rational.* London: Oxford University Press, 1923.

Page, Denys Lionel. *Sappho and Alcaeus: An Introduction to the Study of Ancient Lesbian Poetry.* Oxford: Clarendon Press, 1979.

Plotinus, M. R. *Writings.* Cambridge, MA: Harvard University Press, 1966–1988.

Plutarch. "On Superstition." *Moralia II*; "On Isis and Osiris," "The E at Delphi," "Obsolescence of the Oracles." *Moralia V*; "Dialogue on Love." *Moralia X.* Cambridge, MA: Loeb Classical Library, Harvard University Press, 1928–1996.

Pomeroy, Sarah. *Goddesses, Whores, Wives, and Slaves: Women in Classical Antiquity.* New York: Schocken Books, 1975.

Price, A. W. *Love and Friendship in Plato and Aristotle.* Oxford: Oxford University Press, 1989.

Pseudo Dionysius. "The Divine Names" in Paul Rorem, ed. *Complete Works.* Mahwah, NJ: Paulist Press, 1987.

Rawls, John. *A Brief Inquiry into the Meaning of Sin and Faith*, with "On My Religion," ed. Thomas Nagel. Cambridge, MA: Harvard University Press, 2009.

Richlin, Amy, ed. *Pornography and Representation in Greece and Rome.* New York and Oxford: Oxford University Press, 1992.

Rosenzweig, Rachel. *Worshipping Aphrodite: Art and Cult in Classical Athens.* Ann Arbor: University of Michigan Press, 2004.

Rowe, Christopher. "Socrates and Diotima: Eros, Immortality, and Creativity" in John J. Cleary and Gary M. Gurtler, eds. *Proceedings of the Boston Area Colloquium in Ancient Philosophy.* Leiden: Brill, 1998.

Sandel, Michael. *Liberalism and the Limits of Justice.* New York and London: Cambridge University Press, 1982.

Scarfe, Brian and Christian, Adam, eds. *Beyond Mechanism: Putting Life Back into Biology.* Lanham: Lexington Books, 2013.

Segal, Alan. *Life after Death: A History of the Afterlife in the Religions of the West.* New York: Doubleday, 1989.

Shakespeare, David. *Derrida and Theology.* London: T&T Clark International, 2009

Sheffeld, Frisbee. *Plato's Symposium: The Ethics of Desire.* Oxford: Oxford University Press, 2006.

Simon, Erika. *Festivals of Athens: An Archaeological Commentary.* Madison: University of Wisconsin Press, 2002.

Sissa, Guilia. *Sex and Spirituality in the Ancient World.* New Haven, CT: Yale University Press, 2008.

Shakespeare, David. *Derrida and Theology.* London: T&T Clark International, 2009.

Smolin, Lee. *The Life of the Cosmos.* New York: Oxford University Press, 1997.

———. *The Trouble with Physics: The Rise of String Theory, the Fall of Science, and What Comes Next.* Boston, MA: Houghton Mifflin, 2007.

Spretnak, Charlene. *The Lost Goddesses of Early Greece.* Boston, MA: Beacon Press, 1978.

Tillich, Paul. *The Courage to Be.* New Haven, CT: Yale University Press, 1952.

———. *Systematic Theology.* Chicago: University of Chicago Press, 1967.

Tuana, Nancy, ed. *Feminist Interpretations of Plato.* University Park: Penn State University Press, 1994.

Turner, Scott. *The Tinkerer's Accomplice: How Design Emerges from Life Itself.* Cambridge, MA: Harvard University Press, 2007.

Vernant, Jean-Pierre. "Greek Cults of Aphrodite" in C. Kondoleon and P. C. Segal, eds. *Aphrodite and the Gods of Love.* Boston, MA: Museum of Fine Arts Publications, 2011.

Vlastos, Gregory. "The Individual as Object of Love in Plato" *Platonic Studies.* Princeton, NJ: Princeton University Press, 1973.

———. *Socrates: Ironist and Moral Philosopher.* Ithaca, NY: Cornell University Press, 1991.

Walsh, Carey Ellen. *Exquisite Desire: Religion, the Erotic, and the Song of Songs.* Minneapolis: Fortress Press, 2000.

Whitehead, Alfred North. *Science and the Modern World.* New York: Free Press, 1925.

———. *Process and Reality.* New York: Simon and Schuster, 1979.

Wilken, Robert. *The Christians as the Romans Saw Them.* New Haven, CT: Yale University Press, 1984.

Winkler, John. *The Constraints of Desire.* New York: Routledge, 1990.

Wolfson, Henry. *Philo: Foundations of Religious Philosophy in Judaism, Christianity, and Islam.* Cambridge, MA: Harvard University Press, 1947.

Wolkstein, Diane and Kramer, Stanley. *Inanna Queen of Heaven and Earth: Stories and Hymns from Sumer.* New York: Harper & Row, 1985.

Wolterstorff, Nicholas. *Justice: Rights and Wrongs.* Princeton, NJ: Princeton University Press, 2010.

Wright, Robert. *The Evolution of God.* New York: Little, Brown, 2009.

Zuckert, Catherine. *Plato's Philosophers: The Coherence of the Dialogues.* Chicago: University of Chicago Press, 2009.

Index

GPSR Compliance
The European Union's (EU) General Product Safety Regulation (GPSR) is a set of rules that requires consumer products to be safe and our obligations to ensure this.

If you have any concerns about our products, you can contact us on

ProductSafety@springernature.com

In case Publisher is established outside the EU, the EU authorized representative is:

Springer Nature Customer Service Center GmbH
Europaplatz 3
69115 Heidelberg, Germany

www.ingramcontent.com/pod-product-compliance
Lightning Source LLC
Chambersburg PA
CBHW050046141025
33954CB00004B/380
* 9 7 8 1 1 3 7 5 1 6 0 1 5 *